THE KITTEN OWNER'S MANUAL

Arden Moore

Foreword by John C. Wright, Ph.D.

Storey Publishing

The mission of Storey Publishing is to serve our customers by publishing practical information that encourages personal independence in harmony with the environment.

Edited by Nancy Ringer and Marie Salter
Cover design and art direction by Meredith Maker
Cover photograph © Paulette Braun, Pets by Paulette
Illustrations by Alison Kolesar, except those on pages 168 (aloe, calendula, chamomile, slippery elm, valerian by Beverly Duncan; catnip, ginger, red clover by Sarah Brill; parsley by Brigita Fuhrmann) and 184 by Rick Daskam
Text design by Susan Bernier
Text production by Jennifer Jepson Smith
Indexed by Peggy Holloway/Holloway Indexing Services

Printed in the United States by Banta Corp.
10 9 8

Library of Congress Cataloging-in-Publication Data

Moore, Arden.
The kitten owner's manual / [Arden Moore].
p. cm.
Includes index.
ISBN-13: 978-1-58017-387-2;
ISBN-10: 1-58017-387-X
1. Kittens—Miscellanea. I. Title.

SF447 .M68 2001
636.8'07—dc21

2001032033

To all the kittens in the world
and to the cats that stubbornly remain kittens at heart.
Special mention goes to Callie,
Little Guy (aka "Dude"),
and Murphy.

Contents

Foreword

Cats are easy to keep. They make a home warm and inviting. They gaze at you with understanding eyes when you need a friend. And they're fine with your leaving home for a weekend if you must. But for those who have never shared a home with a cat, entrée into cat guardianship may be an uncertain prospect. Are the stereotypes you've heard about cats true? Are cats totally independent, socially aloof, and untrainable?

Not to worry. Help is at hand. In *The Kitten Owner's Manual,* award-winning author Arden Moore helps even the most "cataphobic" among us to successfully adopt and nurture a kitten through its first eighteen months. Along the way, the author replaces unhelpful stereotypes with fascinating feline facts; throw-the-dice methods of pet selection with personality-based guidelines; and guess-we'll-treat-it-like-a-puppy rearing strategies with developmental accounts of what to expect and how to nurture the perfect feline pet.

Presented in an easy-to-read format are answers to hundreds of questions from perplexed cat owners. Among the book's intriguing topics are tips on deciphering the meaning of kittens' vocal and nonvocal communication; how to decide on the best toys and create the most fun in a safe living environment; the scoop on litter-box success; conventional and holistic health tips; and teaching your kitty tricks, just in case she wants to show off for the dog. Also included are clear, step-by-step instructions for building feline furniture and fun toys, and helpful suggestions on almost anything a new kitten owner could imagine, or not imagine.

The Kitten Owner's Manual may make the novice cat owner more knowledgeable and confident about adopting a first kitten. But be careful. What if you decide to adopt a second? Fortunately, Arden Moore addresses how to introduce a new kitten to your resident cat, too.

— **John C. Wright, Ph.D.**
Certified Applied Animal Behaviorist and
Author of *Is Your Cat Crazy?* (Hungry Minds, 1996)

Preface

Kittens do the darndest things. They tackle toes, ambush ankles, and nip fingertips. They leap like acrobats, race like sprinters, and pounce like linebackers. They wink at you with soft eyes, nestle under chins, and converse in high-pitched mews.

Even when they're "bad," they're delightfully good. It's hard to stay angry at a curious kitten that has just discovered that your flowerpot makes a nifty place to hide her toy mouse. Or a kitten that relishes the panoramic view from atop your refrigerator.

They are our frisky playmates, calming confidantes, and therapeutic lap sitters, all packaged in soft balls of fur. Before you know it, they become cats, adults of the feline society. But many still retain their kitten qualities.

I know. At thirteen, Little Guy, my tiger-striped tabby, still can't resist chasing a shoelace I wiggle in front of his face. Murphy, my one-year-old, seems to study Little Guy's every move and eagerly joins in stalking the shoelace. With Little Guy's gentle guidance, Murphy, the willing student, learns about teamwork and how to wait her turn. Her paw swipes are quicker but his are more on target.

Finding out what makes kittens tick can be tricky. They aren't born with owner's manuals. Too often, we learn as we go — and, as they grow. This book can help. You hold in your hands an insider's guide to kittens. Learn how to counter middle-of-the-night toe ambushes and keep your plants bite-free. Discover why your kitten cackles at crows and purrs like a Mack truck. Tap into feeding tips and litter-box etiquette. Recognize the value of keeping your feline friend indoors and making sure she wears a collar with identification or has an identification microchip.

As a kitten owner, you have lots of questions. This book contains lots of answers. Here's your chance to be as curious as a cat and have fun while you learn.

Paws up!

Acknowledgments

I wish to thank all of the veterinarians, animal behaviorists, animal shelter directors, and cat owners who generously shared their insights into kittenhood to help bring this book together. Special "paws up" gratitude goes to veterinarians Lowell Ackerman, Roger Valentine, and Karen Overall; animal behavior experts John Wright, Larry Lachman, and Kit Jenkins; and Nancy Peterson of the Humane Society of the United States. Finally, I offer special thanks to my editor, Nancy Ringer, for giving me this wonderful chance to share kitty tips with cat fans everywhere.

DEVELOPMENT AND COMMUNICATION

The smallest feline is a masterpiece.

Leonardo da Vinci

Welcome to the Wonder Year. Your young kitten is embarking on a twelve-month journey filled with excitement and exploration, with you as her supportive guide.

Her road to adulthood will be filled with lots of physical and mental growing pains and pleasures. Along the way will be key kittenhood milestones. Understanding how and why your kitten develops the way she does will make your friendship more meaningful and fun.

That's where communication comes in. True, we don't speak Cat and they don't speak People, but a whole lot of communicating goes on with voice tone and body language. Observe your kitten, and listen to what she's telling you.

Developmental Milestones

Kittens may be unpredictable at times when it comes to play, but when it comes to their physical and mental development, they're usually right on target.

Q What happens to a kitten during her first month of life?

A During the first two weeks, a kitten relies on the senses of touch and taste because she can't see or hear. Her mom is her one and only meal ticket and source of warmth. Mom also cleans up the kitten's urine and feces. Around three weeks, a kitty goes from being totally oblivious of her surroundings to realizing that there is a world out there that can be both exciting and risky. A slight competition for rank and turf among littermates begins. Baby teeth start poking through the gums.

Q What occurs during a kitten's second month of life?

A A kitten now possesses all senses and has no trouble finding her mother. By eight weeks, a kitten should be completely weaned from her mother and able to eat solid foods. The kitten's muscles are developed enough to allow her to run, leap, and play with her littermates. You may hear a few high-pitched shrieks as littermates learn how to inhibit their bites during play. A kitten will have sufficient eye-paw coordination to practice prey drives on toys and the dexterity to dance over obstacles. (*Prey drive* is a kitten's built-in hunting instinct.)

Her neuromuscular system is sufficiently strong for her to be able to control her bladder, allowing her to urinate outside the nesting area in the litter box. Kittens also start to groom themselves and others.

Use this time to get your kitty used to having her paws and ears touched and her mouth examined.

Q What type of growth happens between three and seven months?

A Your kitten's permanent teeth will begin coming in during this time, creating a need to chew to relieve gum discomfort. Also, a kitten should reach 80 to 90 percent of her physical size by six months of age. That's what's happening on the outside.

On the inside, your kitten is starting to flex her independence and her desire for individual exploration increases. She wants to test what she has learned by watching her mother and others. This is the prime time for a kitty to scoop, toss, paw, mouth, and hold objects and to initiate tail-chasing and pouncing. It is also known as the *ranking period,* as the relative success of these activities distinguishes between dominant and submissive members of the household.

Her sexual hormones are fully developed. Unless spayed, a female will enter her first heat cycle (a howling period you can definitely avoid by having her sterilized by the time she weighs at least two pounds and is eight weeks of age); this is the first time she can become pregnant. An intact male will want to roam the neighborhood in hopes of becoming a father.

> **FELINE FACT**
>
> Owning a cat is a bargain compared to buying a new car, but make sure your household budget includes the cost of cat care. An indoor cat that lives fourteen years will, on average, cost $6,650 during its lifetime. That price tag covers food, litter, medical care, grooming, and toys.

Q What happens between eight and twelve months?

A A kitten will finish her physical growth by filling out her frame with muscle mass. By her first birthday, she should be 95 percent adultlike in physique, intelligence, and attitude. Experts say complete social maturity usually takes a few more months. This is known as the age of adolescence, when your "teenage" kitten will challenge the household rules and try to test to see how far she can go with her deeds (such as waking you up two hours before your alarm clock rings).

Q What rules do felines live by?

A There are five. Consider these rules as keys to helping you better understand how your kitten thinks.

1. **Felines prefer set routines.** They like to wake up at a certain time, eat at a certain time, and expect you home at a certain time. They quickly learn your daily schedule and adapt accordingly. That may partially explain why some wake us up a few minutes before our alarm clocks chime. It is as if their body clocks sound the signal that it's time to get out of bed and start a new day.
2. **Felines abhor confusion and change.** That explains why some scoot under the bed when your sister's family pays a surprise overnight visit or when they spot packing boxes stacked in your living room.
3. **Felines are turf oriented.** They feel most comfortable inside familiar surroundings. Dogs, on the other hand, are pack animals. Dogs are more willing to accompany their favorite people pals to strange new places. Cats prefer to stay at home. Your home is their castle.
4. **Felines love to sleep.** Some Rip van Felines will snooze up to seventeen hours a day. I've yet to meet a cat with insomnia. They have their favorite napping spots, some sunny, some shaded, some perched up high, and some tucked inside closet corners.
5. **Felines are refreshingly candid.** They never pretend. They never lie. If they don't want to sit on your lap, they will become escape artists and wriggle free. Please don't be offended. At that particular moment, they would just rather be somewhere else. On the other hand, if they want to snuggle next to you, they will boldly stride over and get between you and that mystery novel thriller you can't seem to put down. Without uttering a word, your kitten is saying, "Hey, look at me. Pay attention to *me.*"

Communicating with Your Kitten

In many ways, felines may be savvier linguists than we are. Paws down, kittens read and interpret our body language far better than we do theirs. How else can you explain why your kitty seems to be part psychic and dashes under the bed when you approach to take her to the veterinary clinic? Or how she knows, seconds before you say, "Treat," that this trip to the kitchen will offer a tasty payoff if she tags along?

Kittens are also straight talkers. They never deceive or pretend. Unlike us, cats don't clutter their vocabularies with slang, double meanings, or sarcasm. If your kitten is mad, she'll spit and swat. If she feels playful, she'll be a body of wiggling glee. If she wants solitude, she'll depart to a quiet spot without a lick of guilt.

A cat has emotional honesty. Human beings, for one reason or another, may hide their feelings, but the cat does not.

— Ernest Hemingway

Q My kitty seems to make different sounds at different times. What is she trying to tell me?

A Cats don't need to speak English. Their simple language suits them just fine, thank you. But if you pay close attention, you'll discover that cats communicate clearly through different vocal sounds and body language. Listen and learn.

- **Meow:** Your cat delivers this sound when she demands your attention. She may be saying, "Come see me play — *now!*" Or, "It's late. Where have you been all night?"
- **Chirp:** This musical trilling sound ends in a question mark. It conveys a friendly greeting given only to people, not other cats. It could mean, "Welcome home," or "I just woke up from my nap and it's really good to see you."
- **Purr:** Cats have the unique skill of being able to breathe in and out while making this engine-revving sound with their mouths closed. They purr at twenty-six cycles per second. Cats purr when they are happy — like during a massage — and, strangely, when they confront a stressful situation, such as a car ride to the veterinarian.

- **Murmur:** This short, rhythmical *brrrrp-brrrp* sound is a quick "hi" from your kitten to you. It is often delivered when your kitten is performing a figure eight around your ankles at mealtime.
- **Mew:** This is a pure kitten sound delivered only to her mother. A young kitten will emit this high-pitched squeaky sound in anticipation of being fed.
- **Cackle:** Window-watching felines fixated on birds will make a *ka-ka-ka* machine gun–like noise. Their lower jaw quivers. This is a sound of frustration by a kitten unable to reach her prey: that taunting sparrow on your maple tree branch.
- **Hiss:** Your cat is plainly telling you to "back off." The mouth is open and the teeth are exposed during this verbal warning. Unheeded, the offender may receive a defensive swipe of the paw.
- **Spit:** This short, explosive shout often occurs before or after a hiss. Your kitten is hopping mad and feeling mighty defensive. The message: "If you touch me, I'll hurt you."

Q Why does my kitten purr?

A Experts have yet to pinpoint the purpose of purring, but some suspect that purring is activated by the body's release of feel-good hormones called *endorphins.* These hormones are released in extreme situations, both pleasurable and frightening. That may partly explain why your frightened kitten will suddenly launch into a song of purring at the veterinarian's office.

Q Why does my kitten chat so much?

A Vocalizing is just one way your kitten communicates with you and others. Some breeds, such as Siamese, tend to chat more than others. But if your otherwise quiet kitty sudden turns into a chatty Cathy, she may be in physical pain or discomfort. Take her to your veterinarian to be examined.

If your kitten is otherwise healthy, she may simply be demanding attention. She may have noticed that if she speaks, you will react by feeding her, yelling at her, cuddling her, or talking back to her. Some kittens rescued as strays may talk because they want to go outside. Try to counter this outdoor urge by spaying or neutering your kitten to control her or his hormonal urges. And provide her with a choice windowsill to watch the outdoor action from a safe inside perch. Engage in some play, and distract her attention by hiding toys and treats around the house for her to find.

Finally, some kittens will talk up a storm if they've just lost a favorite person or pet pal due to departure or death. Just like you, a kitten may vocalize to express her grief and sadness. Stick with a routine and give her extra cuddle time and the vocalization should gradually cease.

Q How should I speak to my kitten?

A The happiest kittens feel like part of a family. So speak with your feline friend often. Greet her when you come home. Acknowledge her when you stride past her on the way to the kitchen. Let her know that she matters in your life.

Speak in soothing, flattering tones so she won't feel neglected. And always use your cat's name each time you begin a conversation with her. It helps her recognize her name, and, more importantly, she learns to associate hearing you speak her name with good activities such as receiving praise, getting a friendly head scratch, or being offered tasty treats.

Q Are there any benefits to meowing back at my kitten?

A As silly as it may sound, try mimicking your kitten's vocalizations once in a while. When your kitten meows at you, she may be asking, "Hey, did you see that giant crow feasting on the bird feeder?" Give her a half-wink look and meow back. Don't worry if your meow translates into, "Yes, today *is* Tuesday." Words can get lost in translation, but not your good intentions. Your cat will appreciate even your most feeble attempt at Cat Speak.

Q Can kittens have different personalities, even if they are from the same litter?

A Yes, feline siblings can be as different as your own siblings. Even within the same litter, you can have a few kittens that are shy and withdrawn and a few others that are boisterous and highly energized. Some may be bullies and others big-hearted lovers.

For the most part, cats work out an acceptable living arrangement among themselves without much need for you to intervene. Unlike dogs, cats aren't into hierarchical living arrangements. There usually isn't one "top dog" among the felines living under the same roof. One cat may be in charge of food, another gets to commandeer your lap, and a third may be the keeper of the toys.

> **FELINE FACT**
>
> A litter can range from one to ten kittens with the average litter size being five. The world record belongs to Bluebell, a Persian from South Africa that gave birth to *fourteen* kittens!

Body Language

Kittens tend to "chat" more with their bodies than with their mouths. Let me help you interpret some of the classic kitten communication cues.

Q What are some simple feline body cues I need to understand?

A Unsure of what your kitten is trying to tell you? Here are some consistent cues:

- A kitten that allows you to touch and play with her paws trusts you.
- A kitten flopped on her back with relaxed muscles feels safe and comfortable in her surroundings.
- A kitten that persistently paws your lap is requesting affection. A kitten that insists on jumping in your lap is demonstrating her control over you.

- A kitten that walks up to you, drops her head, and makes forehead-to-forehead contact (known as the *head butt*) feels total affection for you.
- A kitten that tucks her ears flat against her head with dilated pupils feels uneasy about the situation and may fight or flee.
- A kitten that flicks her tongue around her lips is showing signs of anxiety.
- A kitten that begins to twitch her tail, flick her ears back and forth, and act restless is exhibiting the classic "don't pet me anymore" body posturing.

Cats invented self-esteem; there is not an insecure bone in their body.
—Erma Bombeck

Q How does a kitten use her ears to communicate?

A While you should never limit yourself to reading only one body part, the ears can signal important messages. In total, thirty muscles orchestrate movement in each ear of a feline, compared to the mere six that we humans have. These muscles rotate 180 degrees so that a kitten can actually pick up sounds without moving her head. The ears also contain more than 40,000 nerve fibers, which allow cats to hear with ease very high and very low frequencies.

Ears pointed forward usually convey friendly interest or alertness. Ears that twitch convey emotional extremes — pure relaxation or pure tension. Ears that are flat and sideways often mean a kitten's interest has been piqued. Ears pressed flat against the head usually signal fear or submissiveness. Be forewarned: a frightful kitty will attack if she feels she has no escape route.

Q What do my kitten's eyes tell me?

A Eyes are the peepholes to a cat's personality. Let's run down the list:

- **Unwavering stare:** A feline that locks into someone or some animal without blinking is making a challenge.
- **Dilated pupils:** A cat's pupils are light sensitive. They shrink to vertical slits in bright light and expand to dark

pools in darkness in order to see their surroundings. However, back off from a feline with large pupils during daylight: that's a sign of nervousness or agitation.

- **Wide-open eyes:** This eye posture is used when a kitten wants to garner your attention. Use this welcoming opportunity to practice some good behavior. She is a willing pupil at this moment — pardon the pun.
- **Half-open eyes:** Snooze time is just around the corner for this relaxed kitty. It is also a sign of trust if she lets you touch her while she's feeling this relaxed. A kitten that gives you half-open-eyed winks is conveying adulation toward you, so wink back with your sleepy peepers.
- **Closed eyes:** Your kitten is nearing or completely in dreamland. Let her be.

Q What role do whiskers play?

A Jetting out from either side of a feline's face, whiskers are space detectors. If you look carefully, you can count twenty-four whiskers spaced in four rows on each side of your kitten's face. Cats rely on whiskers to solicit information from their surroundings. These whiskers come in handy in tight places. If the whiskers can make it through the width of a narrow opening, a kitten knows that the rest of her body will be able to slither through.

A frightened kitten will press her whiskers against her cheekbones. A relaxed kitten will let her whiskers go straight out to the sides. An irritated kitten will spread out her whiskers and angle them forward.

Q What different messages does a kitty convey with her tail?

A A feline's tail is much like a mood barometer. When the tail is up loosely as she walks toward you, she is signaling confidence and contentment. When she flicks the tip of her tail at you, she is conveying, "Hello, my good pal." A light twitching motion means relaxed alertness.

On the other hand, when your kitten puffs out her tail like a pipe cleaner, she is frightened and scared. If she whips her tail side to side or thumps it repeatedly on the floor, be forewarned: she is irked and angered by something or someone.

A kitten that maintains her tail parallel to the ground while walking is expressing slight interest. A worried kitten will often tuck her tail between her legs.

Q Does a kitten show her emotions with her fur?

A Yes, even a feline's fur can forecast her mood. A cat spooked by the sudden shattering sound of a drinking glass hitting a tile floor will arch her back and puff up her fur so that it stands up on end all over her body. But a truly frightened feline will raise only the fur along the spine and on her tail. She is trying to appear bigger, more menacing, to someone or something that she views as a threat.

Q What is my kitten trying to communicate by rubbing her head against the furniture?

A She is marking her territory, planting the Kitty Flag and declaring the couch, the coffee table, and the recliner as "Property of Me." She does this by secreting chemicals called *pheromones* from scent glands located in her forehead and around her mouth and chin.

Q Why does my kitty rub against my leg?

A A kitten does most of her communicating not with mews but with her body language. If she rubs against your leg — and you know she will — she is marking you. It's her friendly way of telling other scent-skilled animals that, "Hey, this is *mine.*" Don't worry. It's a form of feline flattery.

Q Why do kittens knead with their paws?

A This action, sometimes affectionately referred to as *making biscuits,* mimics the days when your kitten was nursing and used her paws to draw out her mother's milk. Postweaned felines knead when they are happy and content.

Q What does the term *meatloaf position* mean?

A A kitten that feels safe and confident will often sit with her front paws tucked under her belly, hence the meatloaf nickname. With her front legs tucked, it will take a kitten longer to flee an unwanted situation. Conversely, a cautionary kitten will often sit with tensed muscles and front paws out straight so that she can make a quick escape.

Q Why does my kitty sometimes seem standoffish?

A In the words of the late Greta Garbo, "I want to be alone." Nothing personal, mind you. But kittens, just like you, need some solo time to recharge and renew themselves.

So the next time you see your kitten perched on a windowsill or lounging in a favorite sunny spot, don't assume that she's merely being lazy. She may be meditating, tuning out the world and existing in the here and now. My animal behaviorist friends call this the art of silence. When cats do this, they clear their thoughts, reduce their heart rates, drop their blood pressures, and enjoy a sense of calm.

So, instead of disturbing your windowsill kitten, take a cue from her. Treat yourself each day to five to ten minutes of pure, uninterrupted solitude.

Signs of Trouble

Yikes! Your kitty is definitely not in a playful mood. Learning the cause behind this anything-but-pleasant attitude enables you to step in and resolve the situation.

Q What are some signs of stress in cats?

A Your frazzled kitty may act out any or all of the following behaviors:

- Excessive grooming
- Eating less or more than usual
- Suddenly spraying on walls or going outside the litter box
- Acting more passive or aggressive

Q What can cause stress in kittens?

A Kittens don't shoulder the burdens of paying bills or making sure that they get to work on time, but they, too, have their share of stress. Do what you can to minimize stress for your kitten. Many things can put your kitten on edge, but animal behaviorists cite these top stress inducers:

- Making the adjustment from outdoor stray to indoor pet
- Strangers, children, or babies
- Moving to a new home or enduring significant remodeling inside the home
- A dramatic change in your work schedule that keeps you away from home more
- Introducing a new pet to the household
- Barking dogs within view and earshot
- Inability to reach their treasured possessions: food bowl, litter box, and quiet sleeping spots
- Noisy appliances; vacuum cleaners seem to top the list, but air conditioners, furnaces, and smoke detectors can cause panic in some kittens

Q My kitten appears fearful — she is backpedaling, hissing, and hiding. What could cause this behavior?

A The trigger behind your kitten's hiding or defensiveness could be anything. But generally, the cause is a specific person (a stranger, a child, a cat-unfriendly household member), an animal (an aggressive cat or dog), or a loud sound (vacuum cleaners rarely win popularity contests among felines).

Q How can I help my fearful kitten?

A If your kitten suddenly develops a fearful streak, make an appointment with your veterinarian. This behavior may be due to a medical condition. If your kitten receives a clean bill of health, however, you'll need to take a gentle approach.

If your kitten hides each time the doorbell rings, make sure her "safe spot" is always accessible (my Callie feels secure when she can dash under my bed). Let her come out when she's ready. If you try to force her out, she will only become more fearful and reclusive.

You can also slowly build confidence in your scared kitten by sticking to a set routine. A kitten that can set her inner body clock to daily feeding, playing, cuddling, and grooming time will feel more self-confidence.

If you've identified the cause of the fear, you can try to desensitize your kitten. Let's say it's the vacuum cleaner. Leave it unplugged and set it in the living room for a few days. Don't draw a lot of attention to it. Let your kitten approach it on its own. When she sniffs it, praise her. Then start putting a few of your kitten's favorite treats within a couple feet of the vacuum cleaner so that she learns to associate pleasant things with this appliance. Go slow. You may need to take two steps forward and one step back until your kitty develops tolerance of the vacuum.

Instilling Confidence in Your Kitten

One of the best gifts you can give your growing kitten is the quality of self-assurance and confidence.

Q How can I build trust and confidence in my kitten?

A Your kitten studies your every move and can sense how you are feeling. If you present new experiences to her with an air of confidence and support, she will feel more comfortable venturing into these new situations.

For example, your kitty may be extremely curious about what's inside your closed bedroom closet. Invite her to hang around with you and explore your closet while you're deciding what to wear each morning. Then shut the closet before you leave. The mystery of what's behind that door will disappear.

Or, maybe your kitten hates to be shut out of the bathroom when you take a shower or bath. You can hear her mews or insistent paw scratching. Make the bathroom a welcoming place by inviting your kitten inside. Place an old towel or sweatshirt on top of the laundry basket lid or toilet lid as a comfy perch for your nosy cat before you get ready to bathe. Let her sit and enjoy the fragrance of the bubble bath and the moist heat in the air without getting a drop of water on her fur. It's like a cat sauna.

Resist the playful temptation to sprinkle water on your cat. Use this opportunity to strengthen your bond of trust. The next time you need to bring her inside the bathroom to trim her claws, it will be a less harrowing experience.

Q How can I socialize my kitten?

A Socialization is vital to kittens. They need to be exposed to people, dogs, cats, car rides, and household items, such as televisions, vacuum cleaners, and dishwashers, so they can develop awareness and self-confidence.

The prime socialization period for kittens occurs between four and fourteen weeks of age. During this time, your kitten is like putty, ready to be molded in proper kitty etiquette. Introduce your willing pupil to positive experiences and plenty of interactive play with people, cats, and even feline-friendly dogs. This is not a time to make your kitten a recluse. If any of these experiences are absent or unpleasant during this time, a kitty may become apprehensive or adverse to any of them as an adult.

Try to time the arrival of these guests after your kitty's mealtime when she will be feeling full and content. Repetition is key when trying to extend a cat's sociability. Remember, change takes time, so be patient.

Spend time each day massaging your kitty, playing with her toes, stroking her back, and touching her ears and mouth so that she comes to consider these positive experiences.

Q Why is regular playtime important for my kitten?

A Who said learning has to be boring? Kittens can learn plenty during a healthy play session. Set aside at least ten to fifteen minutes twice a day for your kitty. The location for playtime depends on your kitten's personality. If she likes being the center of things, play wherever your family spends most of its time. If she is shy, select a quiet room where you can both play without interruptions.

KITTY CAPER: MEET MOO, WORLD-CLASS SOCK RIDER

One day, I'm getting dressed by the desk and I look down. There is Moo, my black-and-white kitten, commandeering my computer chair — again. She stares in silent fascination as I button my shirt and buckle my pants. I finish dressing and start to take a few steps to get my shoes when I realize that my left foot has become very heavy. I look down. In a silent flash, Moo has vacated her "throne" and is clinging to my sock like she is static electricity. Somehow, she managed to hang on without nicking my skin with her claws. She casts her playful eyes upward as if to say, "What? Why'd you stop? Keep going!"

I oblige and give her a free foot ride. She seems satisfied, and I find myself wearing a slight smile as I head to work.

—Scott Campbell
Washington, D.C.

Help! My Kitten's Driving Me Crazy!

Kittens aren't born with instant manners, but they seem to possess an innate ability to create mayhem, mischief, and madness — all in the name of feline fun.

Nothing is spared by a supercharged, feisty feline in a fired-up frenzy. Leaping, pouncing, climbing, and diving are built into a cat's genetic code. These drive-you-batty antics are normal to a young feline exploring his environment and testing his abilities.

Normal? Yes. Incapable of being changed? No. You can convert your crazed kitty into Mr. Manners (or Miss Manners) by practicing a little kitten psychology. By recognizing what's triggering the misdeed, you will be more successful at addressing the cause rather than simply reacting to the symptoms.

In this chapter, you'll learn how to reclaim your home, divert your kitty's attention away from a misdeed, and encourage acceptable feline behavior.

Five Kitten-Training Commandments

Before we discuss what your kitten might do wrong, let's review what you can do to help direct your kitten's behavior.

1. **Begin proper behavior training as soon as you adopt your kitten.** By starting your kitty off on "the right paw," you'll have a much easier time than you would trying to break bad, old habits. Never reward bad behavior even when it's cute.
2. **Strive for consistency.** Always use the same voice commands and hand gestures so you don't confuse your kitten. For instance, always say "Sit up" and snap your fingers when you want your kitten to stretch up with his weight resting on his hind feet.
3. **Avoid physical punishment.** Your hand should be viewed as a friend, not a foe. Hitting a kitten fosters fear and distrust.
4. **Remember the species.** You own a kitten, not a puppy. The two species have different motivations for what they do. Don't expect your kitten to fetch your slippers. A puppy is a pleaser; a kitten needs to know what's in it for him.
5. **Customize your behavior training to meet the individuality of each kitten in your household.** Some kittens respond better to some techniques than to others.

Deterrents and Distractions

A great approach to stopping a kitten from misbehaving is to distract his attention (that is, to get him to stop what he's doing) and then direct him to a more acceptable behavior. Don't be surprised; you might already have many effective deterrents in your home.

- Double-sided tape
- Aluminum foil
- Citrus-scented sprays
- Can of pennies
- Water-filled spray bottle or squirt gun

Q **What's the best way to deter and distract my kitten from misbehaving?**

A When your kitten starts clawing the arm of your cherished upholstered sofa or chews a gaping hole in your favorite sweater, you need to react quickly — and appropriately. You need to deter and distract him so he ceases the misdeed and performs a more acceptable type of behavior.

Here are some effective ways to stop a misbehaving kitten caught in the act:

- **Sneak up and surprise him.** Felines abhor surprises. While this is only effective during the unwanted activity, most cats will decide that their antics are not worth the bother of dealing with an obnoxious, lying-in-wait human.
- **Try finger snapping.** Most cats will be curious enough to stop what they're doing and pay attention to you.
- **Mimic hissing.** It's the feline version of a verbal warning.
- **Clap your hands together.** But certainly don't applaud their antics.
- **Shake a can of pennies.** Rinse and dry an empty aluminum soda can and insert a few pennies. Give it a shake when your cat starts to misbehave. Cats hate loud noises.
- **Shout "No!" or "Hey!"** Your loud tone should do the trick and make your kitten stop what he is doing.
- **Squirt him with water from a spray bottle or squirt gun.** Keep this water "weapon" handy so you can spritz him while he's in the act. A little water won't hurt him or your surroundings.

Top Kitty Misdemeanors

Certainly, each household presents its own set of circumstances and challenges for a kitten, but following are a dozen of the most common kitty misdemeanors. The good news? Each behavior can be corrected.

- Skipping the litter box
- Clawing the furniture
- Bullying the other pets
- Ambushing ankles and tackling toes
- Yowling nonstop
- Making predawn wakeup calls

- Begging at mealtime
- Bolting outside
- Chewing on electrical cords
- Playing king (or queen) of the countertops
- Drape climbing
- Raiding garbage cans
- Turning tissues and toilet paper into confetti
- Digging in soil and eating houseplants
- Chewing wool

Q Why is my kitty going outside the litter box?

A If your kitten suddenly sidesteps the litter box and defecates on your favorite Persian rug or takes aim at the living room wall, get him to the veterinarian's office for a medical checkup. He may be allergic to the litter or have a urinary blockage.

If no medical problems are detected, your kitty may be acting out because the litter box smells. Scoop out the contents daily. If the behavior continues, offer several types of litters and styles of litter pans to determine what your kitty prefers. And don't overfill the pan. Keep the litter no more than two inches deep. Some kittens detest enclosed litter boxes because while they are using the boxes they become vulnerable to other household pets.

If you keep litter boxes clean, wall spraying and carpet dampening may result from a psychological cause. Your kitten may feel threatened by a new pet in the family, a taunting outdoor feline trespasser, or heightened household stress and be reacting by marking his territory with urine as a feline version of graffiti. The message is unmistakable: "This is *my* turf."

If you catch your kitten backing up against a wall with a quivering tail, calmly walk over, push the tail down at the base with your finger, and distract him with a play activity. If you can't pinpoint the cause of the problem, put your kitten in a large crate with enough room for food, water, blanket, and a small litter pan while you are away from the house. When you're home, keep your kitten's food and water bowls near the targeted spots because cats prefer not to eliminate where they eat.

For more tactics on litter-box warfare, see chapter 5.

Q How can I stop my kitten from shredding my furniture?

A Cats claw to spread their signature scent, shed dead nail tissues, and express anxiety. They aren't purposely seeking out your most expensive — or treasured — sofa or chair to curb their urge.

The solutions are simple: redirect and prevent. If you catch your kitten in the act, startle him by shouting "No scratch!" or by shaking a can with pennies inside. Immediately redirect him to a scratching post sprinkled with catnip. Reward your kitten with treats and praise when he begins to claw. He will quickly figure out that he gets rewarded for working the scratching post and punished for clawing the furniture.

During this transition from sofa to scratching post, improve the likelihood of success by applying double-sided tape to the targeted furniture. Cats hate feeling anything sticky on their paw pads.

Finally, recognize that felines, just like you, want and desire things they can call their own. "Donate" that old chair, buy a durable scratching post, or give your cat a thick log with bark intact for sharpening his claws. Set the log vertically because cats like to stretch upward when they claw.

Q Why is my kitten stalking and attacking my other cats?

A This sudden aggressiveness has many origins. It may be a result of a long-simmering dislike, a traumatic event like a broomstick crashing to the floor (one cat may blame the other), a medical problem, or the addition of a new pet or person to the household.

And a less obvious trigger: an outdoor taunting trespasser. Your indoor kitten could be displacing his aggression for an outdoor cat to the other inside cat. Frustrated that he can't get outside to defend his turf, he unleashes his coiled up anger on his housemate.

Success is never guaranteed, but you can help restore harmony in the household by feeding the feuding felines in separate locations. For intense tangles, try placing each cat in a carrier positioned far apart but still within sight of one another in the living room for thirty minutes at a time. Let them simmer down. Put treats in each carrier. Reintroduce the cats to each other gradually. It's a proven fact: cats can't be happy and mad at the same time.

If you notice your bully kitten dipping his head, shimmering his tail rapidly from side to side, and arching his back end to pounce, distract him with treats or play. But during the midst of a chase, stick to the sidelines and don't interfere, unless there is bloodshed. Make sure each has access to escape routes.

Finally, when both cats are calm or napping, run towels across their coats. Then swap the towels and place them where each cat sleeps and eats to help them get used to the other's scent and make a positive association with the other cat.

Q Ouch! Why does my kitten attack my ankles?

A Sounds like you've got a kitty commando. To your kitten, your ankles represent a tempting target: moving prey. He pounces and wraps his claws around your ankles to hone his developing predatory skills. Many times, he is simply redirecting the need for natural play and his sexual impulses toward you. Yes, even if the kitten is neutered or spayed sexual impulses remain, though they are less intense.

Quite often, the ankle attacks come from males or solo household kittens desperate for some playtime — or playmates. One remedy: get another cat so your kitten can act out his playful prey aggression in a more appropriate manner. Consider devoting more playtime each day to your kitten. Spend five to ten minutes twice a day playing with your kitten with interactive toys such as feather wands or low-power, low-voltage laser lights that cast beams of light on walls and floors for your kitten to chase.

As far as stopping the attacks, try this: Keep a few of your kitten's favorite toys in your pants pocket. Toy mice or small balls are popular picks. When you see your kitten lurking in a doorway, wiggling his back end in anticipation of an ankle attack, toss a toy in front of you for him to pursue. He will be too busy and distracted by this unexpected "invader" to give your ankles any attention. If he happens to make a surprise attack on you before you can toss out a decoy, stop immediately. Your lack of movement takes some thrill out of "the kill." Then toss one of his toys. As he chases the toy, resume walking.

Q How can I stop my cat from grabbing my toes? I need to get sleep at night.

A I share your middle-of-the-night pain. When Murphy first arrived on the scene, she took great delight in pouncing on my toes as they moved ever so slightly under the bedspread, jarring me upright in bed. Now, she sleeps contently at the foot of my bed and leaves my toes alone.

First, understand that it is highly irresistible for a nocturnal, prey-driven creature like a cat to ignore lumps moving under a bedspread. To a feline testing his imagination, this could be an intruding mouse. Let the games begin!

Here are some toe-preserving tactics: Never, ever play "grab the toes under the sheets" when you first wake up. Your kitty will not be able to distinguish between morning and dead-of-the-night toe pouncing. It's a bad habit that can continue well into cat adulthood.

Keep a spray bottle filled with water or a squirt gun on your nightstand. Each time your kitten leaps on your toes, give him a squirt to make him stop. In time, all you'll have to do is just show him the spray bottle and your water-detesting kitty will stop.

Before you go to bed, spend five to ten minutes playing with your kitty to let him unload some of his boundless energy.

As a final resort, ban your kitten from your bedroom by shutting the bedroom door or confining him to another room or crate with bedding and a litter box for the night. Provide a few of his toys that he can occupy himself with while letting you sleep through the night.

Q Why does my kitten howl all the time?

A Your chatty catty may be a born talker — Siamese are known yakkers — or your kitten may be trying to tell you that he doesn't feel well. The causes behind constant meowing are numerous.

Have your kitten examined by your veterinarian to see if the cause is medical. If not, then pay closer attention to the times when your kitten turns up the meow volume. He may be telling you that his litter box is disgusting, or that he hears or smells a cat outside, or that the food bowl is empty.

If you've determined that your feline speaks out of loneliness, leave the radio or television on to keep him company while you're gone. Or, set the radio or television on timers to go off in the late morning.

Q Why do I seem to have an alarm-clock kitty?

A It's difficult to be alert and clear-headed during the predawn hours. Unfortunately, your kitty may be clever enough to take advantage of your sleepiness by jumping on you, pushing up your pillow, pawing the window blinds, or systematically knocking books off a shelf to wake you.

Before you get angry at him, blame yourself. In most cases, owners trigger these early morning wakeups by feeding their kittens first thing in the morning. By getting up and feeding him — or sitting up in bed and yelling at him, you are reinforcing the behavior.

All it takes is a few times for this to become a habit for cats. I know. Callie was my 4:30 A.M. "alarm clock" before I learned to modify her behavior.

So, if you want a good night's sleep, change the habit. Give your kitten his favorite food before heading to bed so he will sleep past dawn. Or, wait an hour or so after you're up to provide the tasty treat. Your kitten will learn to readjust his body clock accordingly.

If you're into electronics, you can position timer devices that shoot out cat treats or toys in an area far from your bedroom so that your nocturnal kitten can play while you snooze.

Return the favor. By that I mean gently nudge your sleeping cat during one of his many daytime snoozes. Engage him in a favorite play activity. Within two weeks, you both should be able to enjoy a full night's sleep because your kitty will be tired by nighttime.

There is no snooze button on a cat that wants breakfast.

—Unknown

Q How can I stop my kitten from begging for table scraps?

A This is a toughie because we often show our love for our pets by treating them to some tidbits off our plates. Mi food, su food.

How can you say no to a pair of emerald eyes and the soft tapping of a friendly paw on your knee as you prepare to take your first bite of food at dinner time? Easy. Try feeding your kitten his favorite cat treat in a different room at the same time. Or, confine your begging friend to an enclosed room far from the dining room until the dishes are cleared from the table.

The best advice: never get into the habit of feeding table scraps, and your kitten won't know what he's missing.

Q I must have an escape artist for a kitten. How can I stop him from trying to bolt outside whenever a door is opened?

A Some kittens seem to emerge from nowhere and attempt to scurry outside whenever you enter or leave the house. It takes super timing and coordination to be able to block his path.

Often an indoor kitten feels a pent-up need to prowl outside because he smells and hears other cats, especially during the breeding seasons. Or, he may just be curious and want to explore the neighborhood.

Try startling your kitten with water from a spray bottle or squirt gun or shake a can of coins, then say "Back" as you prepare to depart. Keep the bottle or can within arm's reach of the door. Or, toss your kitten a treat or toy just before you exit to distract him. Finally, confuse your kitten by randomly choosing different doors to enter and leave. It's impossible for one cat to lay in wait at three different exits.

Q I picked up my phone and the line was dead. Then I discovered why: my kitten had chewed through the cord — again. What can I do?

A Rely on your parenting instinct and childproof (or, in this case, kitty-proof) your home. Consider placing conduit devices over the cords to stop the chewing. Coat the cords with smells the cat detests: hair spray, cayenne pepper, chlorine-based products, or citrus-smelling agents.

Also, try to pinpoint the reason for the chewing. Often, the behavior is a symptom of separation anxiety. Your kitten loves and needs you too much. So, don't make a big deal out of your arrivals or departures. Ignore your kitten for fifteen minutes before you leave and fifteen minutes after you return. Leave a piece of your clothing and a tape recording playing your voice to give your kitten some comfort.

Q How can I stop my kitten from jumping on counters and tables when I'm not home?

A By nature, a cat likes perching on high places. You can satisfy your kitten's instinctive behavior and maintain control of your household by identifying early on which high places are off-limits.

To break your kitten of the habit of leaping on the dining room table or kitchen counters, block off all but a tempting empty space with objects. On the clear space, place double-sided tape, readily available in craft stores and carpet shops. Felines rely on their feet to mark their territories, so they like to keep them impeccably clean. When a kitten jumps on a counter and sticks to the tape, he hates it and decides to get down and search for friendlier high spots.

You can also stop your stubborn countertop climber by placing some cookie sheets on your table or kitchen counter. Add water to them. Then, the next time your kitty leaps up, *splash!* His paws land in an unexpected lake. He will quickly dash out to drier and friendlier perching places.

If you have easy-to-clean kitchen counters, try spraying vegetable spray on them to make them slippery and slick.

These effective strategies work twenty-four hours a day, even when you're away from home. The best part: your kitty won't blame you when he lands on the tape or pan of water. The countertop will lose its appeal and your kitty will look elsewhere for a kinder, gentler perch.

Q How can I stop my kitten from climbing up on my shelves and other high places?

A Cats are height seekers, which explains why they can be found prancing across high bookshelves and tightrope walking on curtain valances.

Meanwhile, we look at their antics with horror and disgust. And, secretly perhaps, we are fearful that their nimble movements will fail them and they will fall and injure themselves.

There are several ways to deter your cat from leaping up to places where he doesn't belong. Success comes from a little preplanning on your part.

First, aluminum foil is an important deterrent to areas you want off-limits to your kitten. Place aluminum foil on counters and shelves. Cats aren't fond of digging their claws into or walking on this shiny foil. Or, choose double-sided tape. Felines hate to get their paws sticky or dirty.

A third weapon: block access. Shut the door to the room where you keep your breakable valuables or fill the shelves chock-full so that there is no paw room.

Make the shelve undesirable by situating a bowl of citrus potpourri or a small eucalyptus potted plant there. Felines turn their noses up at these scents.

Final tool: cave in and give your kitty a safe, high perch he can call his own. Some like the top of a refrigerator or, even better, a windowsill with a panoramic view of the outdoors. Coax your kitty with treats to these sites and provide cushy towels or blankets for him to roost in comfort.

Q How can I save my drapes from my gotta-scale kitten?

A Don't let your kitten's small stature fool you. You've got a mountain climber for a cat. Why do some kittens scale the vertical landscape of curtains and draperies? Because they're there.

It seems that the thicker (and typically the more expensive) the fabric, the better chance that your kitten will dig in and begin his mountainous climb. Following are a few tips that you might find helpful, especially when a spray bottle isn't handy:

- Select tension rods for your curtains so that your kitten's weight will pull them down as he pulls. A few spills will convince him to cease and desist.
- Use the thinnest thread that will hold the drapery when attaching it to the rods. The weight of your kitten will break the threads and make the fabric fall. To reinforce the unpleasantness (and reduce the number of times you'll have to restring the drapes), attach a small tin can with a few pennies inside in an inconspicuous place. It will cause an ear-rattling noise that will make your kitty flee the scene — and, hopefully, not want to return to try another climbing attempt.

Q My kitty is notorious for raiding my kitchen garbage can and spilling trash all over the floor. What can I do?

A Out of sight, out of paw's reach. Stash that trash in a container under the sink and secure the cabinet doors with plastic childproof door latches. They are available at your local hardware store at nominal cost and are a snap to install.

If you don't have the cabinet space under the sink, store trash in a container with a heavy, tight-fitting metal lid that opens like a clam when you push the foot lever down. They are fairly inexpensive and readily available at discount stores.

And take away aromatic temptations that turn your kitten into a feline locksmith. Always toss bones and other trash that could be harmful to your kitty directly in your outside garbage bins. A scavenging kitty could choke on tiny chicken bones.

Q My kitty loves to shred toilet paper and used tissues like confetti. We try to remember to keep the bathroom door closed but forget occasionally. Any backup plans?

A Discourage your kitten from shredding toilet paper with these tactics:

- Balance a plastic cup of water or a small can with pebbles or marbles on top of the roll. This unexpected booby trap may be enough to deter your kitty.
- Fasten a thick rubber band around the toilet roll to keep it from unraveling.
- Install a covered tissue dispenser that hangs more than halfway down over the paper.
- Store spare rolls in cabinets out of paw's reach.

It's also a good idea to keep a small plastic spray bottle filled with bitter apple (available at pet supply stores and drugstores) in your bathroom or wherever marauding kitties might snatch tissues from uncovered wastebaskets. Simply spritz the tissues in the trash can. Cats detest this odor.

Q How can I protect my plants from my kitty?

A If you have houseplants (which should be cat-safe; see chapter 3 for a list of poisonous plants), you can make them kitty-proof from paw digging. Place rocks or gravel into the soil to make digging less tempting. You can also position wooden skewers in a crisscross fashion on top of the soil to prevent your kitten from climbing into the pot and attempting to dig or make a "deposit."

Redirect your kitten's need to dig in the dirt by providing him with his very own kitty garden of greens — grass or parsley in pots. Encourage him to go to these pots when he wants to chew some greens.

Q Why does my kitty suck on my wool clothing?

A Felines that eat wool or other nonfood objects have what's known as *pica syndrome,* which is characterized by a compulsion to eat nonfood items. The cause behind wool-sucking remains a mystery, but animal behaviorists say kittens that have been weaned too young and some breeds, specifically Siamese, Burmese, and Oriental mix breeds, are the most likely candidates. Felines with separation anxiety and obsessive-compulsive behavior also tend to suck and eat wool.

My first cat, Corky, a Siamese, chewed a cannonball-sized hole in my favorite black wool sweater during his kittenhood. Losing that sweater made me become tidier around my house. Now, I keep sweaters safe inside drawers and closets.

If your kitten is eating fabric, first have him examined by your veterinarian to rule out a medical problem. Sometimes wool-sucking is a sign that something is missing from your kitten's diet. Try adding more fiber to his diet — a teaspoon of canned pumpkin is a great source of fiber. Your veterinarian can offer other dietary adjustments.

Feline Aggression

Occasionally, your cuddly kitten may turn into a terror and aggressively start swatting, nipping, and hissing at you — and the other pets in your household. The following discussion should help.

Q What's wrong with my kitten? When she was younger, she was a complete angel. Now she's nearly a year old and has turned into a devil.

A Blame it, in part, to kitty hormones. Kitties entering adolescence (between six and eighteen months) may suddenly forget to use the litter box, attack ankles, or howl to go outside.

Hormones are a major contributor to your kitty's temperament. That's why it is important to have your kitten neutered or spayed before six months. You'll save yourself a lot of grief.

Or, the cause may be a change in the environment. Felines are creatures of routine. Relocations, new members of the household (pets or people), or the loss of a favorite person or pet pal can affect your kitty's mood.

In some cases, a kitten that feels unsafe or unsure may become defensive and want to protect his turf. During times of change and upheaval, try to give your kitty a small room or safe space that he can retreat to whenever he feels a bit anxious.

> **FELINE FACT**
>
> Cats live, on average, fifteen to eighteen years, but some have reached their thirties. The oldest cat on record lived thirty-six years.

Step up your exercise interaction with your teenage kitty. Keep your play sessions to five to ten minutes and make them upbeat and frequent. Exercise offers adolescents the chance to release excess energy in a positive way.

Q Why does my kitty seem to like being petted and then, all of a sudden, turn and nip my hand?

A If your normally mellow kitten sudden turns into a tiger and lashes or bites, it could be due to an injury or a medical condition. He may have a tender spot that your loving hands just touched. It's a knee-jerk reaction to pain. Bring him to your veterinarian to have him checked.

If nothing's wrong medically, most likely a sudden nip or paw swat are simply displays of feline affection. This unexpected nipping or claw-dug-into-your-hand can hurt!

Q How should I react to my kitten's biting or scratching?

A When your kitten bites or scratches you, immediately react with a loud "Ouch!" Do not quickly yank back your hand during an attack because it becomes a moving target. Your kitten will misinterpret this action, thinking you want to continue playing, and chase it.

Kittens in litters let out howls whenever their littermates bite or scratch too hard during play. It's a way of teaching them how to inhibit their playful attacks. Once you yell "Ouch!" stop petting him or playing with him. Give him the cold shoulder. Turn your back and avoid any eye contact. Kittens are social creatures. He will soon learn that going too far in his playfulness will cost him valuable attention from you.

And provide him with a biting outlet. Give him a couple of soft toys that he can sink his teeth into, such as stuffed socks.

Caution: Cat Bites

If your kitty nips you and punctures your skin, go to the sink and run cold water on the wound for a few minutes and wash it well with soap to flush out bacteria. A cat's mouth is filled with bacteria and you don't want to run the risk of developing an infection. Apply some antibiotic ointment to the site after you blot it dry with a towel. See your doctor if the wound worsens and for treatment of deep bites.

Q What preattack signs should I watch for?

A Pay attention to your kitten's tail when you are petting or playing with him. That's a telltale sign of his mood. If the tail twitches quickly side to side in short movements, he is giving you a last-second warning that he is about to attack. Quit while you're ahead and let him be.

Q How can I prevent attacks from ever happening?

A Your kitten takes his play cues from you, so you need to dictate the rules. Never play rough with your kitten by taking your cupped hand and placing it over his face and moving it side to side. He will instinctively fight back with claws and teeth.

Always use hand extensions such as wands or shoelaces when you play with your kitten so that he doesn't view your hand or arm as part of his "prey." And, don't attempt to force playtime on a kitten while he is eating, grooming himself, or sleeping.

Scratching Posts and Pedicures

Cat scratch fever — all kittens are born with the urge to scratch. Fortunately, you can provide them with suitable scratching outlets without sacrificing your sofa.

Q Why does my kitten need to scratch?

A Cats scratch for a variety of reasons. Probably topping the list is pure enjoyment. But scratching is also a cat's prime way to declare his turf. The glands in a kitten's paw release telltale scents that in cat talk translate into "Hey, this is *my* sofa. Keep your mitts off it." But scratching is also a necessary feline form of manicure. Overgrown, neglected claws can get snagged in carpet and cause pain or injury.

So face it: your kitten is going to scratch. Shift your focus toward directing him to more acceptable scratching spots. You'll rescue your furniture and make your kitty happy. It's a win-win situation for both.

The ideal solution: provide your kitten with his own scratching furniture. It can be a scratching post, a nice block of wood, or that old recliner you keep promising that you will toss out. Praise him each and every time he sharpens his claws on *his* furniture. He will quickly learn that the couch belongs to you and that carpeted scratching post belongs exclusively to him.

Q What should I look for in a good scratching post?

A If you're crafty, you can make your own. But there are plenty of good scratching posts available at your local pet supply store. Keep three things in mind when shopping: stability, height, and fabric texture.

Make sure you choose a sturdy post with a heavy, stable base to prevent it from easily tipping over. If it topples or sways under his pull, your kitten won't use it again.

Look for a post that has plenty of height. Remember, your kitten will become a full-grown cat before you know it, so select a model that is tall enough to permit your feline to fully stretch and extend.

Avoid a fabric texture that matches your living room rug or furniture. It can confuse a poor kitty that may not be able to distinguish between scratching places.

Q Where is the best place to position a scratching post?

A Position the scratching post in a location where your kitten likes to hang out. I placed a scratching post between my couch and the basket of kitty toys so that my cats are able to demonstrate their clawing prowess with me as their audience. If you have the space, add other scratching posts in other rooms so when your kitten has the urge to claw, there is a ready target. Prime location: adjacent to where your kitten likes to take afternoon naps. Why? Because the first action most felines perform after a rest is to stand and do a full-body stretch (it brings out the yoga in them).

Q How should I introduce the scratching post to my kitten?

A You need to be your kitten's scratching instructor. Entice him to go to the post by spending a lot of time playing near it with him. Take an old shoelace and run it across the floor and up and around the scratching post for your kitten to pursue and try to grab. Praise him each time he sinks his claws into the post in an attempt to snag the shoelace. You're reinforcing and encouraging him to scratch.

Whatever you do, don't pull up his front paws and go through the scratching motions. Most felines regard this as an insult to their intelligence.

Q What's the best way to trim my kitten's claws?

A Kitty pedicures can be a breeze if you get your feline used to having his paws touched. Gently rub the pads behind the claws. Do this daily for at least a week before attempting a trimming.

When you are ready for a trim session, bring your kitten into an enclosed small room; the bathroom is ideal. You need to eliminate any and all escape routes. Your manicure tool can be a pet clipper or an old toenail clipper. If you use the latter, make sure that you clean it first by dipping it in alcohol and letting it air dry. Then mark it with a tag that identifies it as the kitten's clipper, not yours.

Press gently on the center pad of his paw to extend the claws. You will see a pink section toward the toe joint. This area is called the *quick.* If you accidentally clip this part, your cat will be in pain and bleed. Clip about halfway between the end of the nail and the quick.

Q Does it help to trim my kitten's nails on a regular basis?

A Definitely! Regular pedicures will keep your kitten from damaging furniture due to the need to scratch. Aim for regular trims about twice a month.

FELINE FACT

Count your kitten's toes. Most cats have five toes per front paw and four per back paw. If your kitten has more than those numbers, you've got a *polydactyl* feline. This genetic trait is usually due to an affected male that spreads this characteristic.

Play

Right behind sleeping and eating, playing is a kitten's favorite activity. But a kitten's play is much different from a puppy's. Let's look at how felines have fun.

Q What type of playing should I expect from my kitten?

A Stalking and pouncing are important play behaviors in kittens. Both aid in the proper muscular development necessary for your growing feline. Before you adopt a kitten, stock up on lightweight and movable toys. Best choices include paper wads, small balls that are too big to swallow, and feather wands. By providing a steady supply of these toys, you will give your kitten an acceptable outlet for him to practice pouncing and stalking. This way, your kitten will be less likely to use family members for these activities.

Q What are some interactive games that I can play with my kitty to keep him from misbehaving out of boredom?

A You'll enjoy a closer friendship with your kitten if you spend ten minutes a day engaged in one-on-one play. This play offers many benefits. It will reduce a fear of meeting others, boost self-confidence, bolster muscle tone and coordination, and encourage friendliness in your kitten.

Stumped for ideas? Try some of my favorites:

- **Flashlight Tag:** At nighttime, shine the flashlight beam against the wall of a darkened room and watch your cat take off in hot pursuit. Or, just dim the lights in a room after dinner. Make sure the area has been cat-proofed so that your feline doesn't knock over anything or run into anything as he chases this light beam.
- **Hide-and-Seek:** With your cat next to you, toss a small treat across the room. As your cat zooms after this tasty

prey, slip around the corner out of view and call his name. When he runs to you, reward him with a treat and plenty of praise. Repeat this a few times each day until he gets the idea that it is time to play "find my owner."

■ **Murphy in the Middle:** My youngest cat, Murphy, loves to play before an audience. When people are around, she will rush to her scratching post in the living room and let out a loud *me-ow* as she wrestles a catnip mouse to the ground. That's my cue to solicit a visiting friend to play Murphy's version of cat-chase-the-mouse game. Murphy sits in the middle of the floor. My friend and I sit about ten to twelve feet apart on either end. We toss a catnip mouse so that it just barely clears Murphy's ears. That's Murphy's signal to start leaping and snagging the airborne mouse between us in midflight. Each time she "scores," we lavish her with praise, applaud, and continue this tossing game until Murphy starts grooming herself. That's her way of saying, "Enough fun. It's time to look glamorous."

■ **Follow the Feather:** Take a peacock feather — or even a long shoelace with a toy mouse tied to the end — and run up and down the hallway. Let the feather or shoelace travel on the floor right past your cat. In no time, he'll be up and joining the chase. Once you have mastered the straight hallway route, you can expand to include the twists and turns of your rooms and stairways. Continually call your cat's name and heap on praise with each catch he makes. It's a great aerobic exercise for the both of you.

■ **Kitty in the Bag:** Cats can't resist an open bag on the ground. Within seconds, they've plunged inside and, from the outside, it looks like the bag has come to life. Take a paper supermarket shopping bag and cut off the handles, if there are any. Use a pair of scissors and cut a circle in the bottom of the bag. Attach a toy mouse to the end of a long shoelace. With the bag placed on the floor with the bottom facing you, draw the toy mouse through the circular opening until it reaches about midway inside the bag. Call your kitten so that he is facing the open front end of the bag. Gently wiggle the toy mouse inside the bag and watch your kitty raise his haunches, do the rear-end wiggle, and then dive inside the bag to capture the mouse. Reel the mouse back out of the circular hole. Praise your kitten for her hunting prowess and then repeat these steps.

Kitty Caper: Hair-Raising Playtime

His full name is Bond James Bond and he is full-throttle fun. He's an emerald-eyed, silver-coated Maine Coon kitty that is usually laid back until he decides it's playtime. Then I swear he is a dog in cat's clothing.

His favorite sport is hair-band fetching. I flip them down the hallway like rubber band missiles and he races after them. He fetches them, brings them back, and drops them right in front of my feet. And, as if I don't know what to do next, he lets out an insistent *m-e-o-w* and I fling it again. His twirling meow is loud and continuous until he gets what he wants. But I've learned to tap his "dog" side. Now, when he brings back the hair band, I'm able to tell him "Sit" and he does, patiently waiting until I toss the hair band again.

—Stephanie Baysore
Sharonville, Ohio

Kitten-Proofing Your Home

Wanted: Kitten-friendly home. Must have sunny windowsill perches, a food bowl that never empties, plus scratching posts and cat beds in every room. Stairs optional for nighttime zooming, but hallway runways are a must. Willing to share premises with cat-loving humans with warm laps and a willingness to clean the litter box daily. Sorry, no poisonous plants or aggressive dogs allowed. Call PUR-MEOW and ask for Clipper.

If kittens could describe their ideal home, it might sound very much like this one. But the thought of creating a safe and fun haven for your into-everything kitten may seem daunting. After all, kittens are quick, curious, and stealthy.

The good news is that whether you live in a penthouse, a ranch home, a basement apartment, or a townhouse in the city, suburbs, or country, you can easily and quickly kitten-proof your home without spending a lot of money — and without surrendering your decor to "nouveau feline."

Before Your Kitten Arrives

Caring for your kitten actually begins *before* she arrives in your home. A few days before you plan to adopt a kitten, spend some time kitty-proofing.

Q What can I do to prepare my home for my new kitten?

A Map out your kitty-proofing strategy with these two questions in mind:

1. What do you want to protect *from* your kitten?
2. What do you want to protect your kitten *from?*

Kittens are inquisitive creatures and mischievous by nature, so it's up to us to remove temptations and stow valuables that can be broken. When it comes down to it, kittens don't know the difference in value between your antique crystal flower vase and a paper wad, nor do they really care.

Take a careful stroll through each room of your home. Look high and low. Assume the role of feline bodyguard, and eye your place from a kitten's point of view. Take a room-by-room inventory and try to anticipate what might entice your curious, adventure-seeking feline.

Recognize that your kitten will like high places and insist on walking on shelves in your presence or when you're away. Realize that food left on the kitchen counter (especially tuna or cheese) is too much of a temptation for even the best-behaved feline to ignore.

In short, practice the same parenting skills you've used with toddlers or grandchildren: place childproof latches on doors housing cat food, trash, or other no-cats-allowed items. Tuck electrical cords out of sight, and place breakables and valuables in drawers or behind closed doors. Wrap or cover drapery cords to prevent accidental choking. Don't place stacks of books or magazines where they can easily topple.

Q What can I do to keep some greenery in my home without posing a health risk to my kitten?

A Plants add a feeling of serenity and welcome in the home, but many popular indoor plants are downright deadly to cats. To ensure that your garden is kitty friendly, follow these steps:

1. **Avoid plants that can be poisonous to your kitten.** Failing that, keep them in planters on ceiling hooks far from paw's reach. When situating a plant, think like a cat. Don't place the hooks too near a ledge or other place that a kitten might use as a launching pad to propel herself to these mean greens. *Always* pick up fallen leaves to protect your kitten.
2. **Shut the door.** Did you just receive a beautiful bouquet of fresh-cut flowers to celebrate the arrival of your new pet, some of which are poisonous to your kitten? Keep the flowers in a vase in a room that's off-limits to your curious kitten. Post a note outside the door that alerts others in your household to the fact that the kitten must be kept out of the room and harm's way.
3. **Plan to appease your kitty's "green paw" tendencies by providing her with cat-safe plants.** A tray of edible indoor grass tops the list. Just sprinkle some grass seed over an aluminum pan filled with soil. Place the pan in a sunny spot, and water it to keep the soil moist but not saturated. In no time, grass blades will sprout, creating a dinner salad for your kitten! This mini-lawn will satisfy her need to chew and also reduce the incidence of hairballs. (See page 149 for more on hairballs.)
4. **Grow plants that will benefit kitty.** Stake out a place in your home where you can grow a container of dill or catnip indoors. Dill is Mother Nature's tummy soother, terrific for curbing indigestion. Catnip is a natural stimulant for felines. This herb belongs to the mint family. Wait until your kitten is at least six months

old before introducing her to catnip. A pinch or two of fresh catnip leaves (diced) or dried crumbled leaves on the scratching post jump-starts a kitty's playful nature.

Poisonous Plants

Here's a partial list of plants that can harm cats:

- Aloe vera
- African violet
- American mistletoe
- Asparagus fern
- Azalea
- Baby's breath
- Bird of paradise
- Buttercup
- Clematis
- Cornstalk plant
- Daffodil
- Dieffenbachia
- Easter lily
- Foxglove
- Horse chestnut
- Hyacinth
- Hydrangea
- Iris
- Lily of the valley
- Mistletoe
- Morning glory
- Oleander
- Philodendron
- Poinsettia
- Poison hemlock
- Primrose
- Rhododendron
- Rubber plant
- Tomato vines
- Tulip
- Yew

More than one hundred plants can be toxic to cats of all ages. Ask your veterinarian for a list of poisonous plants native to your area. Signs of possible poisoning include abdominal pain, vomiting, diarrhea, listlessness, muscle tremors, lack of coordination, and fever.

If you suspect poisoning, contact your veterinarian or one of the organizations listed on page 192 for immediate assistance.

Q Besides plants, what other household items might be poisonous to my kitten?

A There are many. They are most often found in your kitchen, bathroom, and garage.

- **Medications:** Aspirin and other medications can be toxic, so don't leave them out loose and be careful where you store them. One tablet of a nonaspirin pain reliever could kill a kitten. Cabinets secured with childproof latches are best.

- **Onions:** Never let kitty sample onions. The chemicals in onions can cause anemia in kittens.
- **Chocolate:** During holidays — and any day, for that matter — don't leave chocolate out for your kitten to sample. Chocolate contains theobromine, an ingredient that can be toxic, even deadly to cats.
- **Household cleaners and other hazards:** Place household cleaners, insect or mouse poisons and traps, and anything else common sense tells you would be dangerous, out of your kitten's sight and reach. Again, cabinets with child-proof latches are best.
- **Bath supplies:** In the bathroom, elevate shampoos, conditioners, soaps, and razors beyond the reach of curious paws and snooping noses.
- **Toilets:** Keep toilet lids down to prevent your kitten from using them as alternative water dishes. The water can harbor bacteria and harmful cleaning chemicals.
- **Antifreeze:** Watch your kitten like a hawk inside your enclosed garage or, better yet, don't invite her in. If your vehicle leaks little droplets of antifreeze and your kitten ingests them, they can damage her kidneys and even cause death. Antifreeze contains ethylene glycol, a sweet-tasting chemical that attracts cats and other animals. My advice is twofold: shoo your kitty out of the garage and to be on the safe side switch to an environmentally safe antifreeze. Select brands containing pylene glycol. It's nontoxic and biodegradable and contains no phosphates. And, it performs just as well as conventional antifreeze inside your car's engine.
- **Paint and paint thinners:** If you are doing home repairs, never let kitty near opened paint cans and paint thinner, and always store them properly. Both are toxic.

Q My veterinarian says I should keep my kitten indoors, but that seems cruel. Why can't I let her roam freely?

A Veterinarians favor indoor living, to help your kitten live a longer, happier life because:

■ Roaming cats face an increased risk of injury and exposure to infectious diseases.

■ Indoor cats live longer than their outdoor counterparts. Humane Society of the United States statistics indicate that the cat that makes her home outside, on average, lives to be five. Cats kept inside can live into their early twenties.

■ Your friendship bond will strengthen if you interact with your cat indoors. Outdoor cats are too busy being predators to give you the time of day. Most view you as free sources of food, water, and, when the weather doesn't agree with them, a dry place to sleep.

Q What about windows? What can I do to keep my kitten safe — and indoors?

A Kittens are true neighborhood busybodies. They are curious and love to lounge on windowsills to check out what's happening outdoors.

Test the tension of your window screens regularly. Metal screens are the most tear resistant. When you're not home, keep windows closed to prevent kitty from clawing at screens and accidentally popping them open and falling out. But leave the blinds open or the curtains drawn back so your kitten bathe in the warm sunshine.

Kitty Emergencies

Try as we may to be protective "parents" to our kittens, accidents do happen. Taking some preventive measures can often eliminate — or at least reduce the likelihood of — kitty mishaps.

Q What should I include in a first-aid kit for my kitten?

A You can't always predict what your kitten may do nor can you always prevent a potential kitty mishap. That's why it's best to think like a Boy Scout and be prepared.

I keep my feline first-aid kit next to my cats' grooming supplies in the laundry room.

You can purchase commercial first-aid kits for cats at pet supply stores or through on-line pet company Web sites. Or, you can save some money by creating your own.

With a bit of luck, you'll never need to use it, but just in case here are the essentials to include in your kit:

- Cotton balls and rolled cotton
- Cotton swabs for the ears and around the eyes
- Antiseptic wipes
- Cold packs
- Nonstick sterile gauze pads
- Lightweight adhesive tape that won't stick to wounds
- Antibiotic ointment
- Rectal thermometer
- Hydrocortisone cream
- Petroleum jelly
- Hydrogen peroxide (3 percent)
- Diphenhydramine (Benadryl) tablets or capsules for stings and bites
- Tweezers
- Styptic pencil or powder
- Tongue depressors
- Penlight
- Scissors
- Ice pack
- Heating pad
- Plastic eyedropper or syringe
- Mineral oil to remove tar and other sticky materials
- A clean towel
- An old pillowcase to confine your cat during treatment
- Latex gloves
- Phone numbers of your veterinarian and emergency pet clinic

Learn Kitty CPR

Equally important to keeping a well-stocked kit is learning some basic pet first aid. Contact your local humane society or American Society for the Prevention of Cruelty to Animals (ASPCA) shelter to locate a pet first-aid class in your area. Instructors will teach you how to perform kitty CPR, stop choking episodes, take a temperature, treat a minor wound, and other vital lessons.

Q We live in North Carolina, a hurricane haven. And we have relatives who live in Texas, which is prone to tornadoes and hail, and California, which is always susceptible to earthquakes, wildfires, and flash floods. All of us own cats. How can we prepare for a natural disaster and protect our cats?

A We don't always have advance warning about severe weather, and often we have only minutes, even seconds, to react. Designate a place in your home for your kitty first-aid kit and a kitty-survival kit. Store both in a location that is dry and not in direct sunlight.

For the survival kit, include these items:

- An extra breakaway collar with an identification tag that includes your phone number and your veterinarian's number
- A photo of your kitten placed inside a plastic bag that you can show to people in case your kitty flees or gets separated during a natural disaster
- An extra harness and leash
- One week's supply of dry food stored in airtight, waterproof containers
- One week's supply of canned food with peel-back lids
- Plastic spoon or knife to mix food
- Canister of cat treats
- One gallon of water in a plastic jug

- Water bowl
- One week's supply of scoopable litter, a litter box, a handful of plastic bags for disposal, and a new plastic scooper
- A roll of paper towels
- A small, plastic bottle of antibacterial soap
- A few of your kitty's favorite toys and one of your old, unwashed T-shirts for comfort

Store all of these essentials inside an airline-approved cat carrier. Each January and July replace the food and water with a fresh supply.

Need Disaster Relief?

United Animal Nations, a nonprofit group, operates the Emergency Animal Rescue Service, which provides first aid and shelters in disaster-struck areas. This group also works to reunite pets with owners.

The American Humane Association, ASPCA, and Humane Society of the United States also provide disaster-relief programs.

If your kitten becomes lost during a disaster, contact PetFinders. See Resources (page 192) for contact information.

Safeguarding Your Stuff

Don't let your kitten turn into a cat burglar. She won't pawn your valuables, but she will certainly "paw" them if given an opportunity.

Q My kitty loves to jump on my nightstand and bat off my earrings. Why is she so intrigued by jewelry?

A Kittens are attracted to small, shiny objects, such as jewelry, coins, paper clips, and needles. Unfortunately, they can swallow these objects and choke or injure internal organs.

Never leave a pin, needle, or buttons on a table after sewing. The same goes for paper clips, thumbtacks, rubber bands, and other tiny office supplies. Make these items off-limits to your kitten by stashing them inside drawers or containers. If you're a dedicated flosser, protect your kitten by discarding the used dental floss in a wastebasket with a lid.

Q Is it true that kittens can damage bedding and clothing?

A Yes. Some kittens can and do chew or claw holes in fabric. To discourage this behavior:

- Keep a fitted sheet on the underside of your bed to prevent your cat from eating the synthetic materials on the bottom of the box spring. It can keep your kitten from harm and save wear and tear on your box spring.
- Stow wool and knit blankets in a trunk or behind closed closet doors when not in use. Cats that eat these fabrics and other nonfood objects often suffer from pica syndrome, a condition not fully understood by experts. The ingested materials can cause obstruction of the intestinal tract and may be potentially toxic.
- Finally, if you iron regularly, train yourself to stow the ironing board and unplug the iron as soon as you're done. An ironing board offers an unsteady perch for an active kitten, which can accidentally cause a hot iron to tumble, possibly burning kitty and starting a house fire.

Q I spend a lot of time inside my den working on my computer. How can I protect my computer from kitty hair?

A As a kitten, my Callie loved to leap on top of my monitor for an evening snooze while I tapped away on the keyboard. The top of the monitor emits heat, so it's almost irresistible to kitties. Unfortunately, Callie's hair spilled inside my monitor, which required that I get it cleaned. Eventually I bribed her away from this unsteady perch by placing a plush cat bed next to me on the floor of my den for her evening catnaps.

Protect your cat and your computer with these simple safety precautions:

- Choose computer hardware and accessories that can bear a cat's weight. Avoid cantilevered trays and components with fragile, unsupported attachments.
- Place keyboards and other fragile components under shelves to reduce your cat's "pounce potential."
- Rely on printers and scanners with internal paper trays and enclosed feed mechanisms, or cover the feed rollers.
- Always turn your printer or scanner off when you leave the room. One wrong paw can create quite a paper trail.
- Don't block the computer's cooling vents. Locate outlet strips in ventilated areas.
- Use a mouse pad with a smooth surface that easily wipes clean of cat hair. Also, protect your keyboard with a flexible plastic keyboard cover when not in use.
- Resist the urge to let your kitty climb into your lap as you type. I've known computer-whiz felines that have tapped the ESCAPE button and even CONTROL-ALT-DELETE series, causing work to be lost.

Q What's the best way to keep electrical cords out of my chew-happy kitty's mouth?

A Contrary to popular belief, most felines ignore electrical cords. But if you happen to have that rare exception, you need to block access. Chewing live wires can cause mouth burns and, worse, electrocution.

Tack electrical cords to baseboards (be sure not to insert a tack directly into a wire) or wrap them in protective plastic to keep them from being snakelike temptations. Try inserting the wires into plastic decorator shower-rod covers that you can cut to size. Spritz cords with bitter apple spray to discourage chewing, or coat the cords with other smells cats abhor: hair spray, cayenne pepper, or citrus-scented agents.

Q How can we kitty-proof our Christmas tree?

A Follow these simple steps to keep Yule time from turning into yowl time.

- Don't place tinsel on the tree. Your kitten may be attracted to the tinsel's shine, giving it a taste and possibly choking on it.
- Stick with nonbreakable ornaments made of wood, cloth, fiber, or wicker. Glass ornaments can shatter to the floor and cut your kitten's footpads.
- Cover the water inside the tree stand with aluminum foil to prevent your kitten from trying to quench her thirst. Stop your kitty from becoming a mini-mountaineer and attempting to scale your live tree by spraying the lower branches with bitter apple, eucalyptus, lemon, or other feline-foul scents. Alternatively, opt for an artificial tree that lacks that great outdoors smell kitties crave.

Q The holidays are such a busy time. How can we make sure our home is safe for our kitten during all the activity?

A Ensure that the holidays remain happy and healthy by taking precautions to protect your kitten from food poisoning and forbidden goodies. Dispose of garbage in a secure container out of your kitten's reach. Here are some other tips:

- Resist giving your kitten table scraps. Something as common as bread dough can cause tummy aches, bloating, and vomiting in a kitten.
- Don't give your kitten chocolates. If you leave goodies out for others to enjoy, be sure to keep them in sealed tins or jars or containers.
- Watch where you place lighted candles. Keep them covered and out of paw's — and a flicking tail's — reach.

■ If you must have a mistletoe kissing ball for Christmas, bring it out for a quick kiss and then store it inside a closed drawer. Mistletoe is poisonous.
■ Finally, don't invite your kitten to share your holiday cheer. An ounce of alcohol can cause poisoning and death.

Coexisting with Kitty

If you think about it, your kitten actually spends *more* time in your home than you do. She deserves some "squatter's rights," so take a few extra steps to ensure that your house feels like home to her, too. Content kittens are happy and well behaved.

Q How can I coexist peacefully with my kitten?

A Kittens have feelings, too. They need to feel like they belong and that they are valued members of the household. Translation: give them their own furniture. You can claim the couch and recliner; they have dibs on the scratching post and top bookshelf.

■ **Scratching post:** One is a must. Two or more in different rooms are even better. Cats with claws need a place to hone their nails, mark their territories, and release their predatory aggressiveness. The multitiered ones are more expensive, but they double as kitty play forts.
■ **Big soft pillows or folded blankets:** Situate these comfy items strategically throughout your home. Cats like them on the foot of your bed, near a sunny window, and even in a closet. Make sure these sleeping materials are washable and launder them once a month.
■ **Fabric window platform:** Are your windowsills narrow? No problem. Consider investing in fabric platforms that attach to window frames. My cats enjoy the perches that attach to the wall, right below the windowsill, with suction cups. The shelves are sturdy and safe, and you don't have to drill any holes in the drywall.

■ **Grooming supplies:** Cats are fastidious about cleanliness but can always use your help. Keep your grooming kit (comb, brush, cat nail clippers, cat shampoo, and deflea shampoo) in one location, such as in a bucket in the laundry room or in a bathroom cabinet.

Purr-fect Palace

In San Diego, California, Beauregard, Frank-the-Friendly-Kitten, and the rest of the feline gang live inside a three-bedroom ranch house that is literally the cat's meow. Each room outdoes the previous one in terms of feline appeal. There are rugs with smiling cat faces on wooden floors. Toys everywhere. Floor-to-ceiling scratching posts. Cat-sized holes in the upper walls near the ceiling, with brightly colored catwalks cascading and weaving down to the floor.

Owners Bob and Francis Walker have willingly catered to their cats' needs, spoiling them with more than 100 feet of catwalks, a cats-only clubhouse, and window perches for afternoon siestas.

Bob offers this explanation: "If possession is nine-tenths of the law, then our place is truly the cats' house. They spend more time there than we do. We believe our cats are safer living indoors and need a place that piques their curiosity and makes them feel safe."

And, the Walkers are willing to share their knowledge. Each year, they sponsor an open house for curious visitors, most of them cat lovers. The Walkers donate the price of admissions to the National Cat Protection Society in the hope that other cat owners will pick up some pointers on how to make their homes more feline-friendly.

Q Sometimes my kitten seems to disappear in the house. Where can she be hiding?

A These tiny Houdinis are drawn to tiny, cozy, dark, out-of-the-way places. Favorite secret spots include behind a refrigerator or stove or inside a clothes dryer. Solution: block access to these dangerous appliances. When not using the washer or dryer, always keep the lids shut. And, silly as it may sound, always check inside these appliances for a wayward kitty before using them. While you're at it, arrange furniture so your kitten can't get stuck behind or inside anything.

Also, make sure that your kitten isn't snoozing inside a dresser drawer, cupboard, or closet when you close the door. When first-time visitors come to my home, I always know where to find Little Guy. He ducks into the bottom shelf of an antique dresser given to me by my grandmother. All I see on top of my folded shorts and T-shirts is a pair of nervous wide green eyes staring back at me. The door is broken, so he can enter and exit as he pleases.

A final favorite hiding spot: under a recliner with the footrest extended. Before you leap into the chair to watch your favorite program, lift the footrest and check to see if your kitten is hiding under there.

FELINE FACT

Kittens sleep, on average, about seventeen hours each day. If you're into math teasers, that means that by the time your kitten is seven years old, she will have slept away about five years of her life.

Let the Fun Begin

Your kitten spends many hours home alone while you're at work, but that time doesn't have to be filled with naps and boredom. You can set up some fun games to keep your kitten highly amused and occupied while she is the sole occupant of your house.

Q Okay, my house has been kitty-proofed. Now, what can I do to keep my kitten entertained when she's home alone?

A Cats may spend up to seventeen hours a day snoozing. That's an average. Some active cats love to play most of the day. While you're at work, keep your cat occupied with toys and activities. Not only are they entertaining, but toys help recharge a lethargic cat and help temper the energy level of a frantic feline. Here is a rundown of some of my cats' favorite homemade playthings:

- **Cat's in the bag:** Place a brown paper shopping bag on its side on the floor of your living room or dining room. Be sure to cut off the handles so your cat won't accidentally choke. Now, just before you head out for work, sprinkle a teaspoon of fresh or dried catnip inside the bag, deep in the bag. Your cat's super scenting ability will drive her right into the bag for fun. At the end of the day, a sweep of the broom or a quick vacuuming will clean up the mess in seconds.
- **Cat swat toy:** Take an old shoelace from a pair of sneakers. Tie one of your cat's favorite toys on one end and wrap the other end around an interior doorknob so that the toy dangles about four or five inches from the floor. Most cats can't resist walking by this toy without giving it a good swat.
- **Sock it to 'em:** Fill an old cotton sock with tissue paper and a pinch of dried catnip leaves. Tie the open end of the sock into a knot. Give the sock to your kitten and watch her bat it around.
- **Nothing to sneeze at:** Think of this homemade toy as a Rubik's Cube for kittens. Take an empty tissue box and place a Ping-Pong ball inside. Your kitten will spend hours trying to fish out this rolling sphere from the narrow opening. It's a real cat teaser, and it gives new life to an old tissue box.
- **Kitty treasure hunt:** Before you leave for work in the morning, take five or six of your cat's favorite toys — catnip mice, paper wads, shoelaces, whatever — and hide them around your house. Hide a few treats, too. Great hiding spots include under the couch, behind a pillow, and on a windowsill. Play this game with your cat a few times so that she gets the idea. Then, once she is ready to go solo, praise her when you come home for the booty

that she finds. My cats like to dump their found loot next to the scratching post in the living room.

- **Cardboard bed:** Cats like nothing better than to cozy up inside a small area. Place a medium-sized cardboard box on the floor when you leave. When you get home, don't be surprised if your cat is snoozing inside or has tucked some of her favorite toys inside the box. My cat Callie lives for cardboard boxes. Separated from her mother at two weeks old and found wandering the streets of Miami, Callie was never weaned. So, she loves to chew on the cardboard box and spit out the pieces, littering the floor. My vet has checked her teeth (tartar-free) and says this is a harmless pastime for a slightly neurotic cat.
- **A true fish tale:** Cats can spend hours watching fish weave back and forth inside an aquarium tank. Make sure that the aquarium's lid is securely attached to avoid any cat-pawing episodes. And place the aquarium on a sturdy stand so it can't be tipped over.
- **Run a tub tub:** Toss a plastic practice golf ball inside your bathtub for your kitten to hone her sprints and quick turns in a confined game of chase.
- **Light and sound show:** Have your lights and radio set on timers so that your cat will hear sounds and see lights coming on and off to make her feel more at ease.
- **Shake, rattle, and roll:** Fill empty plastic medicine bottles (with childproof caps) with dried beans or peas. Snap on the lid, call your kitty's name, give the bottle a few good shakes, and roll it across the floor. The noise and the motion will prove irresistible to your little hunter.
- **All keyed up:** Got a handful of old keys? Put them on a key chain and dangle them just above your kitten's head. Felines can't say no to shiny, metallic objects. Encourage your kitten to test her balance by resting her weight on her hind legs as she tries her best to bat the keys with her front paws.

Making a Kitty Tetherball

Kittens like to jab and swing like prizefighters. Let them unleash these urges in a constructive way by making tetherball toys. You can buy one of these clip-on toys or make your own. The lightness of the wire allows the fabric ball to move erratically, drawing the curiosity of your kitty predator.

Materials	**Equipment**
Fabric	Heavy-duty thread
Cotton batting	Needle
22-gauge steel wire	Wire cutters
Large plastic clip	Needle-nose pliers

1. Fashion a fabric ball and stuff it with cotton batting.
2. Cut a 3-foot length of steel wire. Attach one end of the wire to the fabric ball with heavy-duty thread. Slip the other end of the wire through the hole in the handle of the large plastic clip. Use needle-nose pliers to fasten the wire securely and to tuck in wire ends.
3. Connect the clip to a doorjamb or on the edge of a sturdy piece of furniture.

Making a Scratching Post

This scratching post is inexpensive, easy to make, and durable. Your kitten will thank you for one and will be even happier with two or three.

Materials

Thick piece of plywood, at least 24 inches square
18-inch length of thickest closet-rod dowel
2 sturdy angle irons with fitting screws
Small carpet remnant
Heavy hemp rope or sisal rope
Nail

Equipment

Sandpaper
Soft marking pencil
Small power drill
Screwdriver
Contact glue
Contact cement
Hammer

1. Sand and smooth out the edges of the plywood to remove all splinters.
2. Use a pencil to mark the center of the plywood, and place the dowel upright on the mark.
3. Set the angle irons on opposite sides of the dowel. Mark the placement of screw holes on both the dowel and the board.
4. Drill small holes to start the screws, and screw the angle irons tightly in place.
5. Cover the board with a small piece of carpet, gluing firmly in place.
6. Apply contact cement to the dowel, beginning at the bottom. Work on small areas at a time.
7. Once the cement has dried, wrap the dowel tightly with heavy hemp rope, pushing each spiral close to the previous one. Add more glue to the post as you work upward.
8. Finish the top with a tight single knot and nail in place, so the knot is on the top of the post. Cut the rope a few inches above the knot and unravel the end to make a stiff brushlike tassel for paw swatting.

Q My kitty loves scratching posts. How can I choose a good one for her?

A As mentioned earlier, kittens need furniture that they can declare: Mine, Mine, Mine. Save your prized sofa by buying or building a few scratching posts for your claw-happy cat.

A few ground rules, however, before you begin your shopping trip. First, select a post that provides both vertical and horizontal scratching areas for your growing kitten. It should be tall enough for her to fully stretch and extend her upper body and front legs from a sitting position.

Second, scrutinize the covering. If you opt for carpet remnants from your local carpet store, go with a high-quality grade so that the scratching post will last longer.

Third, shop for stability. You want a post that is sufficiently sturdy so it won't topple over when your kitten stretches and sinks in her claws.

Toys to Avoid

These toys are unsafe for your kitten:

- Plastic bags with handles
- Soft foam balls that shred easily
- Toys with itty-bitty parts or glued-on pieces that can be swallowed
- Empty cellophane cigarette wrappers that can cause choking

Q Is there a way for me to treat my indoor kitty to the great outdoors safely?

A Indoor kittens deserve to smell fresh air and feel warm sunshine on their coats. If your kitty isn't fond of walking on a leash (see page 184), make the great outdoors safer by creating a pet enclosure that, in essence, will give your kitten her own private room with a view.

Protected inside a sturdy enclosure equipped with scratching posts, ramps, and perches, your kitten has the chance to get a closer look at birds, squirrels, and other outdoor critters without the perils of being out there on her own. And, she can even catch bugs!

Scheduling just ten minutes a day for your kitten to romp and snooze inside one of these enclosures can invigorate her and help chase away any boredom blues. Be close by as she enjoys spending time checking the outdoor world from a safe location.

Here are a few tips to help you get started:

- Check with your city or homeowner's association first to make sure the enclosure won't require a special building permit or violate rules.
- Be realistic about the size and shape. The enclosure should be big enough for your cat to move about easily, but it needn't match the size of your existing patio or living room.
- Do a budget to estimate the costs in advance. Include the cost of equipment, materials, and time.
- Select quality materials that are sturdy and will last. Best choices include plywood, redwood, PVC piping, thick chicken steel wire, and 4 x 4 blocks of wood. Floor options range from grass and dirt to concrete and carpet. Finally, rely on quality fasteners to prevent accidental openings.
- Take your time building the enclosure or hire someone handy to build it for you.

For folks living in apartments, consider patio-style enclosures that fit over double or single windows, giving your cat her very own bay window area. Some friends of mine have opted to put their cats in large collapsible steel crates on their balconies. The crates fold up neatly when not in use.

Regardless of the size or style of the enclosure, make it comfy for your cats by including bedding materials (thick towels or a cat bed), food, and water. Other options include shelving and connecting tunnel systems.

A final option: if you have an enclosed run in your backyard for your dog, then periodically let your cat use this safe outdoor haven. What should you do with your dog? Use this time to give him a bath with your garden hose. This way you can keep an eye on both your pets.

Kitty Caper: Hide, You Seek

Lager, my grayish brown tiger-striped kitten, loves to play his kitty version of hide-and-seek. He hides his favorite green ball and expects me to find it. It started the day I bought this ball from the local pet supply store. It's the size of a golf ball but soft and spongy.

Lager leaped on it right away and batted it out of my view. Then he came to me, gave me a series of meows, and I followed him. It was almost as if he was testing me on my hunting abilities. Now, when I come home after work each day, it's a ritual. Where is the newest hiding spot for his green ball? Lager has "hid" the ball in his water bowl, in my work boot, and under my pillow.

— Kevin Moore
Washington, D.C.

Kitty Cuisine and Calisthenics

Food is fuel. That adage holds true for you, your kitty, and other living creatures. Providing your fast-growing feline friend with the right foods in the right amounts will help him grow up to be a strong and healthy cat.

Choosing the Best Food for Your Kitten

During the first few weeks of life, your kitten's one and only meal ticket is his momma. By four weeks, his tiny digestive system should be ready to handle bits of semi-solid moistened food. That's when you take over as your kitten's personal chef. This is your golden opportunity to introduce your kitten to healthy eating habits.

Q What must my kitten eat to have a nutritionally balanced diet?

A Between the ages of six weeks and nine months, your kitten should be fed a high-quality diet specifically designed for kittens. Kittens need a minimum of forty-one nutrients. Some of the headliners in this group include amino acids, fats, vitamins, and minerals.

Quantity aside, the nutrients must be delivered in the correct proportion and form to maintain your kitten's health. Commercially available growth foods for kittens accomplish both of these objectives.

Don't feed your kitten a popular homemade "weaning" recipe consisting of a mix of milk, baby cereal, vitamins, eggs, and meat. Not only is this recipe pricey and time-consuming to make, it's not nutritionally complete or balanced.

Kitten Feeding Timetable

0–4 weeks	Mother cat nurses her litter.
4–6 weeks	Kitten continues to nurse but can also eat small amounts of moistened kitten food.
6–8 weeks	Mother cat gradually weans kitten; kitten should be exclusively fed moistened commercial kitten food.
12–14 weeks	The kitten's teeth have developed sufficiently to handle dry food.
First year	Kittens should be fed kitten food for all of the first year.

Q How can I tell if the food I'm feeding my kitten meets his nutritional needs?

A Unfortunately, our kittens can't simply tell us: "How about a little more protein in my diet? My coat isn't as thick as it should be," or "That bargain-priced bag of kitten chow doesn't cut it. My tummy is upset all the time."

Because kittens can't verbalize their dismay, it's up to us to look for signs of protein deficiency in their diets. Don't worry. It's easier than you might think. Kittens served poor-quality food (ones low in digestibility and protein) will display potbellies, experience frequent bouts of diarrhea, be susceptible to infections, and fail to reach developmental milestones on schedule.

Other telltale signs can be found in your kitten's skin and coat. If your kitten isn't getting enough protein, his skin

Understanding the Digestive Process

If you're curious about what happens to your kitten's food when he bites, chews, and swallows, here is a quick behind-the-scenes summary: The kitten grasps food with his teeth and lips and either chews the pieces or swallows the bites whole. Inside your kitten's mouth, saliva moistens and lubricates the food so it's more easily swallowed. The kitten's tongue then pushes food back through the pharynx into the esophagus, the all-important tube that connects the mouth to the stomach. Food immediately drops down the esophagus into the stomach.

Once food reaches the stomach, it becomes saturated with gastric juices. Glands secrete hydrochloric acid and digestive enzymes, which initiate the food breakdown process. Once food has been mixed and treated with the acid and enzymes, it passes into the intestinal tract for final digestion and absorption.

becomes thinner and more susceptible to bacterial infection. His coat will feel brittle to the touch and, in some spots, the hair may be thin or sparse.

Knowing your kitten's nutritional needs helps you evaluate and choose the appropriate cat food. Select brands that contain at least 30 percent protein.

Q Okay, so what should I feed my kitten?

A You can't go wrong if you feed your growing kitten a brand-name food; this is no time to save a few pennies and buy generic brands. Good brands are Eukanuba, Iams, and Science Diet. Each offers specific dry and canned foods scientifically prepared to meet the nutritional needs of kittens.

When selecting a brand of kitten food, make sure that the name and address of the manufacturer is listed on the label. Reputable companies also list toll-free numbers for consumers to call for additional information or to ask nutritional questions.

The label should indicate that the diet is adequate for its intended purpose. You know you're buying quality if an animal source of protein is listed as one of the first two ingredients on a canned food label and as one of the top three ingredients on a dry food label. The order of ingredients indicates the relative proportion of ingredients. Lower-quality brands contain less protein and so list it much lower.

Also, pet foods labeled as COMPLETE and BALANCED bear the seal of approval from the American Association of Feed Control Officials (AAFCO). This group sets the standard for the minimal vitamins and minerals required to meet the nutritional needs of kittens. The nutrients are carefully balanced to provide just the right amount of nutrition for your kitten.

Finally, feed kitten food until your kitten reaches his first birthday. He may look like an adult, but he needs a full twelve months of kitten diet.

Q Should I serve my kitten low-fat food?

A No. Kittens need fat to keep their coats and skin healthy and to provide them with energy. In fact, one ounce of fat delivers twice the energy of one ounce of protein or one ounce of carbohydrates. Fat also helps make the food more palatable.

Kittens need essential fatty acids (EFAs) to keep their skin and coat looking and feeling its best. Unlike dogs, kittens also require *arachidonic acid,* a fatty acid that is commonly found in fats from animals, poultry, and fish. Be sure that the kitten food you choose contains at least 9 percent fat, with a small amount of arachidonic acid. This hard-to-pronounce nutrient is essential for our feline friends.

Q What's the best way to introduce a new type of commercial food to my kitten?

A When you first bring your kitten home, keep him on the same brand of food for a few days to a week. After that, if you want to switch to a different brand, gradually mix the new food with the old food to help avoid digestive upsets, food refusals, or diarrhea that may occur if his diet is changed too abruptly.

Here's a handy chart:

Day(s)	Old Brand	New Brand
1–3	75%	25%
4–6	50%	50%
7–9	25%	75%
10	0%	100%

Q Which type of food is better for my kitten, dry or canned?

A Actually, both are acceptable, providing you buy national brands from reputable food manufacturers. Let's run down the pros and cons of each:

- **Dry food:** You'll spend the least and get the most for dry food. It can be left all day in your kitten's bowl without risk of spoiling. Most brands are nutritious and they provide that needed crunch that will help keep your kitten's teeth strong. Kittens have sharp little teeth, and by six to eight weeks of age they can chew dry food just fine.
- **Canned food:** Aaah, the aroma is irresistible — canned food draws most kittens to their bowls like magnets. But it comes with a price: it costs more. The average high-quality, 3.5-ounce can costs a bit less than a dollar. After opening, and providing you seal it with a lid, canned food can be safely stored in the refrigerator for a few days; ideally, it should be used within three days of being opened. It also must be refrigerated to be kept fresh. Be forewarned: if you feed your kitten only canned food, you run the risk of creating a fussy cat that

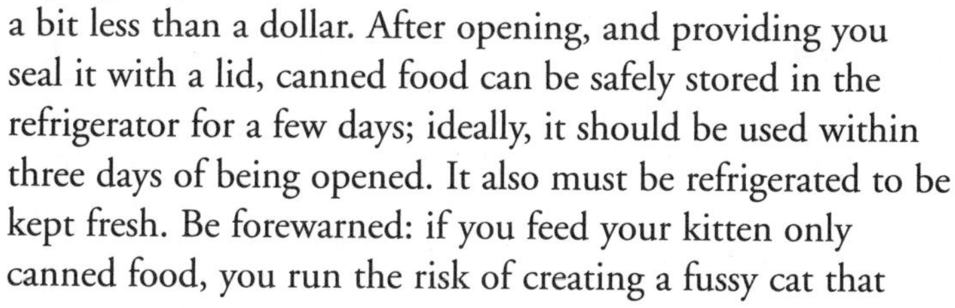

will paw aside dry food. Instead, try serving dry food as the main meal and use the canned food as a treat.

After your cat has eaten, remove canned food left in the food bowl and wash the bowl in hot, soapy water to prevent the growth of bacteria, such as *Eschirichia coli (E. coli)* or salmonella. Once a week, put your kitty's food and water dishes in the dishwasher or wash them well by hand.

Food Shelf Life

Dry kitten food stored in an airtight container will stay fresh for up to one year from the date of manufacture. After that, the chemicals in the food ingredients start to break down, causing the food to become rancid. Canned food is good for up to two years when left unopened.

Q Which minerals are important for the health of my kitten?

A Knowing your kitten's nutritional needs will help you evaluate and choose the appropriate kitten food. Minerals essential in the feline diet include iron, copper, potassium, zinc, manganese, selenium, and iodine, says Lowell Ackerman, a veterinarian and expert on cat nutrition from Mesa, Arizona.

Let's run down the role for each:

- **Iron:** Found inside red blood cells; helps carry oxygen from the lungs to the rest of the body.
- **Copper:** Assists red blood cell function and connective tissue metabolism.
- **Potassium:** Helps body fluid, nerve transmission, and certain metabolic processes.
- **Zinc:** Essential to normal skin, bone, muscle, and hair growth.
- **Manganese:** Vital for normal cell reproduction and blood clotting.

- **Selenium:** Required, along with vitamin E, to prevent oxidation damage in the body by free-radical molecules.
- **Iodine:** Works in harmony with thyroxine, a hormone produced by the thyroid gland that regulates the basal metabolic rate.

Another good clue that you're buying quality food? Reputable brands list trace mineral salts toward the bottom of the list of their food ingredients. If you see manganous oxide, ferrous sulfate, sodium selenite, and calcium iodate listed, the diet contains adequate amounts of these minerals, says Dr. Ackerman.

Way down deep, we're all motivated by the same urges. Cats have the courage to live by them.

—Jim Davis, Garfield cartoonist

The Cat Café

If you enjoy cooking and occasionally want to treat your young cat to some homemade meals, you're in luck. The Cat Café is open for business. Allow me to share a few recipes that are truly the cat's meow. Be sure to put leftovers in plastic containers with lids and store them in the refrigerator. They will stay fresh for about one week if refrigerated.

Supurr-b Scrambled Eggs

Makes 2 servings

1 tablespoon of margarine
2 eggs, beaten
¼ cup cottage cheese

Melt the margarine in a small frying pan over medium heat. Add the eggs and cottage cheese and stir until the ingredients are well mixed — usually about 2 to 3 minutes. Pour half of this recipe into your cat's bowl and allow it to cool before serving. Store the other half in the refrigerator and serve it the next day.

Tuna-Rice Cats-erole

Makes 6 servings

1 cup of chicken broth (low-sodium)
pinch of oregano
½ cup green beans, finely chopped
½ cup uncooked, instant brown rice
1 hard-cooked egg, finely chopped
1 six-ounce can of tuna (water-based)

In a medium saucepan, bring the chicken broth to a boil, then reduce heat and add the oregano, green beans, and rice. Cover and allow to simmer for 10 to 15 minutes or until the rice is cooked. Add the egg and tuna and stir well. Serve at room temperature.

Colossal Cat Chowder

Makes 6 servings

½ pound white fish, deboned and diced into small cubes
1 cup creamed corn
1 cup skim milk
¼ cup red potato, finely chopped
1 tablespoon liver, finely chopped
1 clove of garlic, minced
pinch of salt
¼ cup low-fat grated cheese

In a medium saucepan, combine all the ingredients except for the cheese. Cover and simmer over low heat for 20 minutes, stirring occasionally. Remove from heat and sprinkle with cheese. Allow this dish to cool to room temperature.

Heavenly Kitty Hash

Makes 4 servings

- 1 cup water
- ⅓ cup uncooked brown rice
- 2 teaspoons corn oil
- pinch of salt
- ⅔ cup lean ground turkey
- 2 tablespoons chopped liver
- 1 tablespoon bonemeal

In a medium saucepan, bring the water to a boil. Stir in the rice, corn oil, and salt, and reduce the heat to low. Allow the mixture to simmer for 20 minutes, covered. Then add the turkey, liver, and bonemeal. Stir frequently and simmer for 20 minutes. Cool to room temperature before serving.

Wonderful Water

Water is vital to your kitten's health. Water helps the kitten regulate temperature, digest food, transport salt and other electrolytes through the body, and eliminate waste. Keep your kitten's water bowl clean. Fill it with fresh spring or filtered water several times daily.

FELINE FACT

Water constitutes up to 84 percent of the body weight of newborn kittens and up to 60 percent of the weight of adult cats.

Q My kitten barely touches his water, even though it's out all day long. What can I do?

A If your kitten has boycotted the water bowl for a few days, be on the safe side and contact your veterinarian. Your kitten may have a medical condition that needs attention. If medical concerns are ruled out, dab a few drops of water on your kitten's mouth several times a day. He will automatically lick it. If your kitty just doesn't seem to be a big fan of water, sneak in some of his daily water needs by serving canned food, which contains a high percentage of water.

Q I fill the water bowl every day, but my kitten likes drinking from the faucet instead. Why?

A Many kitten behaviors can't be explained, and this is one of them. One theory holds that the sense of taste in cats is stronger than that in people. Therefore, some kittens insist on quenching their thirst by drinking dripping water from a faucet because they prefer fresh-tasting water. Some cats prefer a water bowl right next to their food dish. Others must be part-raccoon: they seem to enjoy dipping their front paws into their water bowls and licking off the moisture. Generally, it doesn't matter how a kitten gets his water, as long as he gets enough from a healthy source. But be sure to keep toilet lids down and bathroom doors closed. The chemicals and germs in toilet water can make your kitten sick.

Q Is it okay for me to give my kitten milk?

A Experts give a conditional yes. Some kittens develop milk, or lactose, intolerance and experience gastrointestinal upset (vomiting, diarrhea) when allowed to lap up milk. If you share a home with such a kitten, keep the milk for yourself and for your growing children. If your kitten can tolerate an occasional milk treat, serve skim milk and never chocolate milk. Chocolate contains theobromine, an ingredient that is toxic to cats.

Q I want my kitten to have strong bones. Is it safe to feed him calcium-rich yogurt?

A Yes, as long as your kitten can digest the yogurt without displaying any signs of lactose intolerance, such as diarrhea.

Obviously, felines diagnosed with lactose intolerance should not eat yogurt or any other dairy product.

As a yogurt fan, you recognize that it contains healthy amounts of calcium as well as protein, carbohydrates, and fat. Yogurt is a bone-builder for both you and your kitty.

Just don't go overboard on the yogurt servings or your kitten will grow up to become a fat cat. Limit the yogurt treats to no more than one tablespoon per day of the low-fat variety, and make sure you provide your kitten with a high-quality kitten chow that offers a balanced amount of nutrients. By the way, yogurt is about 80 percent water, so your kitten is getting hydrated while she licks up her dairy treat.

Feeding Schedules and Routines

Ah, to be a growing kitten without having to worry about that dreaded word: diet. Still, you need to instill good eating habits in your kitten, so start early and be consistent.

Q How often should I feed my kitten?

A Kittens between six weeks and six months of age should not be restricted in the amount of food they consume. Excess caloric intake, excessive growth rate, and obesity are not problems in growing kittens. Experts recommend that you feed kittens at least three times per day or give them free access to as much kitten chow as they want. Cut back to two meals per day once your kitten is six months old.

Resist sharing table scraps during this period because you run the risk of creating nutritional imbalances and finicky eating habits. Be strong! I know it's hard. It's important for your growing kitten to receive well-balanced nutrition offered in commercial kitten foods.

FELINE FACT

During the first five months, a kitten can experience a 2,000 percent increase over its birth weight — literally growing right before your eyes! Kittens also require about twice the calories per pound of body weight as an adult cat.

So, how much is the right amount? Fortunately, quality brand-name kitten food list serving guidelines, by kitten weight, on their packages. Weighing your kitten regularly will also help you detect any unusual weight gains or losses.

As a general rule, kittens between 7 and 9 pounds need about 8 ounces of dry food or 6 to 8 ounces of canned food per day. The amount of food needed depends on the individual cat and the nutrient density of the food. Also keep in mind that a kitten's appetite and total food consumption may vary slightly from day to day. This is normal.

Q My dog gulps down his food, but my kitten likes to nibble throughout the day. Is that normal?

A Perfectly. Cats tend to prefer eating small snacklike meals throughout the day rather than saddling up to the food bowl and ravaging its contents like an all-you-can-eat buffet. Nibbling is healthier and easier on your kitty's digestive system. Just make sure that your kitty's food bowl is out of paw's reach of your hungry hound.

Q My kitten has always been a healthy eater, but lately, food doesn't seem to interest him. What's wrong?

A Your kitten may stop eating because of an illness or an injury, or even because he's feeling depressed. Monitoring your kitten's eating habits will help you help him, should the need arise.

- If he hasn't eaten all day, have your kitten examined by a veterinarian. Don't wait — go immediately. During the physical exam, the veterinarian will ask you to identify your kitten's symptoms while trying to locate the site of the problem. The veterinarian may try to relieve your kitten's symptoms and improve his health by administering fluids, antibiotics, appetite stimulants, or an anti-inflammatory drug. If your kitten does not respond to such supportive care within a few days, tell the veterinarian, who will then perform further diagnostic tests to assess your kitten's condition.

- Felines with fevers often lose their appetites and their friskiness. If you suspect your kitten has a fever, bring him to your veterinarian. Don't reach for a baby aspirin, ibuprofen, acetaminophen, or other over-the-counter medications — all are highly toxic to cats.
- Kittens are creatures of habit. Sometimes, a kitten won't want to eat because the location of his food dish keeps changing. Try to serve his meals in the same place at basically the same times.
- During hot weather, your kitten's appetite may diminish. This is normal.
- Finally, check with the rest of the members of your household. They may be feeding your kitten lots of treats and table scraps without your knowledge, which would explain why he doesn't want to rush to the food dish when you place it down.

FELINE FACT

Want to venture a guess as to how many mouthfuls of food your cat consumes per day? The answer: twelve to twenty, depending on the appetite of your feline feeder.

Q We have one kitten and two adult cats. Any suggestions on how we can conduct a peaceful mealtime?

A Feed your adult cats separately from your kitten — in different rooms. This will prevent your adult cats from eating your kitten's chow and your kitten from eating food that his young body is not yet ready to handle. Until your kitten reaches a year old, maintain this feeding routine. After that, you can free-feed the cats in one place because all will be eating adult food.

Table Scraps and Finicky Felines

Kittens quickly learn that what you're eating seems to smell, look, and probably taste much better than the food in their bowls, so be careful.

Q Is there any harm in giving my kitten some table scraps?

A Most of us cat owners can't resist begging eyes and occasional begging paws when we sit down to eat. But table scraps don't provide the complete and balanced diet that your kitten gets in commercial cat food.

Tiny pieces (about the size of your thumbnail) of broiled chicken, turkey, or fish are okay to offer as occasional treats to your kitten, but not before he's six months old. Before six months, his digestive and immune systems have not developed sufficiently to handle these people foods.

Never give a kitten raw meat, fish, or poultry because it can harbor bacteria or parasites. Also, never give your kitten any foods containing onions. The chemicals inside onions can destroy your kitten's red blood cells, causing anemia.

Q Help! I've got a kitten that begs each night at dinner. I can't seem to say no to his imploring eyes. What can I do?

A Our feline friends quickly learn where the food is and in no time flat are intrigued by your dinner plate. Resist the temptation to feed your young feline unhealthy table scraps or during your mealtime. If you do, your kitten may be at increased risk for developing obesity and other chronic diseases as he gets older, not to mention that you're cultivating a bad habit: begging. Here are some strategies to help you.

- **Establish a feeding schedule.** Kittens are habitual creatures and enjoy eating at regular intervals. They don't wear little watches, but they have an uncanny way of knowing when it's time to eat. Also, try to serve the food in the same location every day.
- **Feed your kitten *before* you sit down to eat.** Ideally, your kitten will finish eating before you are ready to start. If his belly is full, he probably won't be as anxious to try to beg from you.

- **Feed your kitten in a separate room, preferably one with a door.** Get your kitten started on his food, then close the door behind you. This will allow you to enjoy your meal without interruption, and you can take pleasure in knowing that your kitten is enjoying an undisturbed, proper, nutritious diet.

If you try all of these techniques and still find yourself surrendering to begging eyes, exercise some discipline. Hold off on giving your kitten a people-food treat until *after* you've finished your meal and put the dishes in the sink. Then walk over and place the treat in his food bowl. He'll quickly catch on that he won't get any table scraps at your mealtime and stop the begging.

FELINE FACT

Think you've got a big cat? Your tubby tabby is nothing compared to the late — and weighty — Himmy. This Australian cat earned an entry in the Guinness Book of Records back in 1986 for tipping the scales at a whopping 45 pounds.

Q How many are too many treats?

A Your kitten can pack on the pounds if you go overboard in handing out tasty treats. Moderation is key. Snacks should never exceed 10 to 20 percent of your kitten's total diet. If you want to do the math, experts suggest that healthy cats should take in about 250 calories of food per day. So, the treats should never total more than 25 calories per day.

Q My kitten loves raw tuna but turns her nose up at canned tuna. Is she a tuna snob?

A Felines prefer one food over another based on taste and palatability; the two are inseparable. Cats are sensitive to the taste, texture, and physical shape of their food. These factors determine a food's feline palatability, and palatability is crucial to a cat's decision to eat something. Studies indicate that, not surprisingly, cats prefer a new or novel shape or flavor of food to the same old flavor, day in and day out.

So, if you've always offered him canned tuna and as a rare treat, you offered him a bit of raw fresh tuna, your kitten was probably attracted to this morsel's new shape, strong fishy smell, and taste. Also, raw, fresh tuna acts as a natural toothbrush for cats. Chewing on a piece of tuna massages your kitty's gums and takes away surface tartar — a good way to save on dental bills down the road!

It's worth noting that veterinarians who specialize in nutrition advise *against* giving kittens a steady diet of tuna. Tuna contains magnesium, a mineral that, in high quantities, can cause stones to form in a kitten's bladder. Give tuna only as a treat, not as your kitten's mainstay meal, and you'll sidestep any danger.

Q Is it safe to give my kitten raw shrimp?

A Your kitten's digestive system cannot tolerate raw shellfish. Raw shrimp may contain parasites. And cats with thiamine deficiencies may suffer seizures, even die from eating raw shrimp. Better to be safe than sorry — don't give your kitten raw shrimp. Bake it first.

Q I heard that eggs help keep a cat's coat looking healthy. Can I give my cat raw eggs?

A No. Lowell Ackerman, a veterinarian from Mesa, Arizona, who is a nutritional expert on feline foods says that raw egg whites interfere with the absorption of biotin, one of the important B vitamins. And raw eggs may also harbor salmonella. If you want to dish up some eggs for your kitten, feed no more than one or two cooked eggs per week.

Mealtime Problems

It's inevitable. At some point in their lives, all kittens will have eating problems. The following strategies should help.

Q Why is my otherwise-sane kitten eating dirt from my potted plant?

A This strange food fetish is not unusual. Some felines eat dirt because they are searching for trace minerals missing from their diet. To be on the safe side, have your kitten examined by your veterinarian. A blood profile will rule out any serious medical conditions such as anemia. If your kitten gets a clean bill of health, rest assured that her dirt eating won't harm her. Attribute her actions to one of many feline quirks. Discourage this behavior by covering the soil with aluminum foil or rocks.

Q I'm concerned about the lining in food cans, especially when there is leftover food. How safe is it?

A Canned cat food does not age like fine wine, especially after it has been opened. It should be used within three days after being opened. That said, some owners are concerned about the can's metal lining, fearing perhaps that flecks of metal may get into the food and be swallowed by their kittens. Rest easy. Pet food manufacturers generally use cans with an inert lining, a protective coating. As long as you keep a lid on the opened can and store the can in the refrigerator, the contents will be safe and highly edible for your kitten.

As every cat owner knows, nobody owns a cat.

— Ellen Perry Berkeley

Q I try to keep a clean house, but periodically we get invaded by ants. How can I prevent ants from reaching my kitten's food bowl?

A Here's an age-old ant secret: they can't swim. You can stop an army of ants from commandeering your kitten's bowl of food by setting the bowl in a slightly larger bowl of water, creating a watery moat. Ants won't risk drowning to reach your kitten's kibble.

Q Yikes! My adorable kitty has gas. How can we clear the air?

A First, check the condition of the foods you're serving. Foods that are spoiled are likely to cause a kitten to emit pungent gases, such as methane. Grains, legumes (including soybeans), as well as dairy products can also trigger episodes of gas. Always keep your kitten's food supply fresh, and store it in airtight containers.

When you're shopping, read labels and select kitten foods that are highly digestible, low in fiber, and contain moderate amounts of protein and fat. Consider serving small meals several times a day instead of one large meal. Mini-meals are easier on a kitten's digestive system. Some kittens gulp food down and swallow air — another trigger for gas release.

If, after all of these steps, your kitten's gas still persists, have your veterinarian take a look at your kitty for other possible causes, such as food allergies or an intestinal condition.

Q My kitten turns up his whiskers at most foods. How can I make his meals more tempting?

A You need to regain control of mealtime. Once your veterinarian has ruled out a medical cause, start by setting a limit for chow time for your finicky feline. If you have a finicky eater, you need to intervene and set some mealtime controls. If he doesn't finish eating within an hour, pick up his food bowl and put it away. Later in the day, set it down again for an hour, and then pick it up. Your kitty will catch on quickly that he better eat when the food is served or he'll go hungry for a while.

Here are other helpful ways to make your kitten a member of the clean-the-bowl club:

- **Serve it warm.** If your kitten snubs the bowl of dry food, add a little warm water. When mixed with the food, water makes a light gravy, just the thing that might make the food more appetizing. Try warming up canned food in a microwave-safe container for ten to fifteen seconds. Test the food's temperature with your finger before serving; it should be warm but not hot.

- **Bring on the fishy smell.** In the world of kittens, the smellier the food, the tastier it is. Treat your kitten to some fish-flavored vittles and watch him turn into a chowhound.

> **FELINE FACT**
>
> Kittens can detect sour, bitter, and salty tastes but cannot detect sweetness. A kitten's tongue is also designed to detect food textures and temperatures.

Q My kitty is a sloppy eater. Any tips to keep cleanup to a minimum?

A Some kittens will take mouthfuls of food out of their dishes and scatter the food all over the floor. They are simply doing what comes naturally. Some move away from the food source to prevent other cats from taking food away from them.

Break this habit by making sure that your kitten is allowed to eat in peace without any hassle from other family pets. Place his food bowls in an area out of heavy foot traffic, too.

If you have a cat that spills a lot of food near the bowls, try serving the food and water in heavy dishes that won't slide around. Also, place plastic mats or newspaper under the dishes to keep any spills from traveling to the rest of the floor.

Kitty Calisthenics

Congratulations! You've become the controller of your kitty's cuisine. Equally important is maintaining a regular workout routine for your kitten so he doesn't become fat. Here are some fun exercises. See chapters 2 and 3 for more fun games.

Q I realize that in my own life, healthy eating and exercise work together in keeping me fit. Any advice on how I can keep my kitten in shape?

A Kittens love to play, so turn play sessions into calorie-burning workouts. Here are a few fun kitten exercises.

- Use an interactive toy that has feathers on the end. Swish it up, down, and sideways to mimic a bird's movement and to encourage your kitten to leap and stretch.
- Create a kitty obstacle course in your home by setting up bags, cardboard boxes, and soft-sided tunnels in a room. Watch your kitty leap and zoom.
- Slither snakelike toys up and down stairs, and kitty will chase after them.
- Invest in a multitiered cat scratching post so that your kitty can leap from one perch to the next. Your kitten will have a blast while honing his balance and coordination skills.
- Play kitty hockey. Toss a colorful plastic ring from a milk jug on a noncarpeted floor and test your kitty's speed and coordination. After playtime, remember to pick up the "puck" — you don't want your kitten to accidentally chew and swallow a small piece of plastic.

Kitty Caper: Thirsty Tina

When I rescued Tina from a local humane society, she was four months old. Little did I know just how clever she would prove to be. A few days after she arrived, I awoke to the sound of running water. The kitchen faucet was on. I thought my father or I had forgotten to turn it off , but it kept happening.

Then I caught Tina in the act. She jumped on the counter, turned the water handle with her paws until the water was running, then drank from the running water. When her thirst was quenched, she jumped down from the counter. I came up with a solution accidentally. One night I left a glass of water in the sink. Tina didn't turn the water on that night. So, when I remember to leave a full glass of water in the sink, the faucet stays off. But any night I forget, Tina is back working her paw magic.

— Nancy Sachse
Cedarburg, Wisc.

Litter-Box Lessons

"Kitty, allow me to introduce you to the litter box — your personal place to potty." If only it were that easy!

The truth is, kittens need your help to learn proper bathroom etiquette. Sure, they may be born with a predatory drive to pounce on fast-moving paper wads and to cackle at birds, but they need to be schooled in the finer points of litter-box use. Some learn it from their mommas or littermates; others turn to you as their teacher.

Understanding your kitten's needs from the get-go will help make her homecoming smooth and uneventful.

Litter-Box Basics

True, shopping for litter boxes and accessories may not be the highlight of your weekend, but selecting the proper items will help your kitten develop healthy habits.

Q My kitty arrives tomorrow. What supplies do I need?

A Litter 101 class begins before your kitten steps foot in your home. You need to buy the right bathroom accessories and essentials for your kitten. The shopping list includes a plastic litter pan (easy to clean), litter, a plastic durable pooper-scooper, a dust pan and broom, antibacterial liquid cleaner, and an air freshener. Bypass plastic liners — they are a hassle to use and a needless expense.

If you are adopting a tiny orphan kitten under eight weeks of age, you'll need a small plastic box with sides no higher than three inches. The sides should be high enough to keep the litter inside and low enough for your short-legged feline to easily step up and over.

For kittens older than eight weeks, purchase a standard-sized plastic litter box (18 inches long by 14 inches wide by 4 inches high). Kittens are able to spring in and out of the litter box with their limber legs. And, trust me, they will because at this tender age everything is a game or an adventure. Some kittens like covered litter boxes for the privacy, and owners like the way they help contain the mess and odor. But if you have a kitten that is easily intimidated by other cats or household pets, place an open litter box in a corner so that your kitten can see and prepare for any unexpected interruptions.

Q I have three kittens. How many litter boxes do I need?

A Three. This is my favorite math lesson. One box per cat should be your goal. And, if possible, locate the litter boxes in different places in your home. This strategy will help discourage one kitten from being territorial and blocking access to a litter box by another feline housemate. It's physically impossible for a territorial cat to stand guard in front of all of the litter boxes at the same time. And, having an extra litter box or two provides easy access and reduces the likelihood of "mistakes."

Q What type of litter should I use?

A When it comes to selecting litter, you have many choices. Types include clay, clumping, and organic (recycled newspaper). You make the call for your kitten. Continue to use the same type unless she starts to avoid her bathroom box, a clue that she may be allergic to the litter or prefer something different.

Litter Tip

When you first bring a new kitten into your home, use the same brand of litter she was used to in her previous home. This will help make your kitten's arrival to your home a more familiar transition. If you prefer a different brand, you can gradually add it to the litter box and reduce the original type of litter once your kitten has mastered the bathroom routine.

Q Where's the best place in the house to keep the litter box?

A Let's start by ruling out the worst places. Topping the list: right next to your kitten's food and water bowls. Felines have super-strong noses that can detect aromas far better than people can. And they are into the clean scene, too. So make sure there is distance between the food and water bowls and the litter box. Another no-no locale: a high-traffic area of the home such as the kitchen, dining room, or living room. Cats prefer a little privacy and quiet when they do their business.

So, the best places: in a bathroom, the closet floor of a spare bedroom, a warm cozy corner of the basement, or in the laundry room, providing that the litter box is not right next to the vibrating washer and dryer. By placing the litter box in a good location, you will encourage good litter-box habits in your kitten. Strategically place an air freshener about five feet away to help temper litter-box odor.

Litter Box Do's and Don'ts

As you begin teaching your kitten how to use her litter box, keep these tips in mind:

- *Do* keep your kitten's litter box clean by scooping out the deposits daily.
- *Do* provide an individual litter box for each feline member of your household.
- *Do* pay attention to your kitten's litter preferences. Does she prefer the clumping type or the sand type?
- *Do* locate the litter box in a secluded, quiet area.
- *Don't* use perfumed litters.
- *Don't* locate the litter box near food or water bowls.
- *Don't* use ammonia-based cleaners on the litter box. The scent is too strong and unpleasant to most felines.
- *Don't* allow dogs, children, or other cats to deny your kitten access to her litter box.
- *Don't* medicate, groom, or do anything else that the kitten perceives as unpleasant in or around the litter box.
- *Don't* ever rub the cat's nose in urine or stool if it misses the box. Punishment only confuses the kitten and may make the situation worse.

Q What's the proper way to introduce my kitten to her litter box?

A Kittens are natural copycats. Kitty see, kitty do. Many learn the appropriate place for urination and defecation by observing and imitating their mothers. If your kitten has been orphaned and there is an adult cat in your household, she may pick up pointers from the adult cat, especially if she has been accepted as a new feline friend. The adult cat becomes the kitten's litter-box tutor.

If your new kitten has no other feline companions, you must become the instructor. Don't panic — it's easier to teach a kitten how to use a litter box than it is to teach your child how to ride a bike. When your kitten nibbles on food or laps up some water and starts to walk away, this is your cue. Wait about fifteen to twenty minutes, then slowly walk over, pick up your kitten, and place her in the litter box. A kitten's bladder is tiny and the digestive process is fairly quick, so this is prime time for litter-box action in a kitten.

When your kitten is in the box, spark her interest in the litter by stirring it gently with your finger. Let the kitten jump in and out of the litter box at will. Don't force her to stay in the litter box. The goal is to make the litter box a place of necessity — one without unpleasant associations or tension. Once your kitty makes a deposit, give her a few gentle strokes and offer praise in a calm, friendly tone.

Most kittens catch on quickly. Within a few days, your kitten will be strutting over to the litter box all by herself.

Be careful not to let the kitten wander too freely at first. She might get lost in the house and choose an inappropriate litter-box area out of desperation.

Q My new kitten seems scared of her own shadow. How can I introduce the litter box to her?

A New kittens often hide until they feel comfortable in their surroundings. Place the litter box where your kitten feels safe, and gradually move the box, if necessary, if she wants to explore the house. At first, keep your kitten confined to a limited area with an accessible box until she is fully litter-box trained. And don't forget that good behavior deserves rewards. Always give her gentle verbal praise for a job well done. She doesn't need food treats during her bathroom training.

Q How often should I clean the litter box?

A Daily. If you get into the habit of scooping out your cat's deposits each morning, you greatly reduce the likelihood that your kitten will seek potty targets outside the litter box. The popular clumping style of litter needs to be scooped daily to keep it fresh. Clay and organic litter types need to be scooped daily and totally dumped and replaced once or twice a week.

Ideally, once a week you should clean out the litter, clean the pan with mild detergent and hot water, rinse it thoroughly, and dry it completely. But once a month will suffice. Be sure to clean the plastic pooper-scooper at the same time.

You can save some money if you use plastic bags from the supermarket to gather your kitten's daily deposits rather than buying cat litter bags. Seal them shut and toss them into your outdoor garbage to keep your home odor-free.

Q Is it true that kittens sometimes sleep in their litter boxes?

A File this under the "Strange but True" category. Some kittens, including my nothing-embarasses-me Murphy, like to nap in their freshly cleaned litter boxes. To them, the box represents a security blanket, providing them comfort. This is more common among newcomers who are adjusting to unfamiliar surroundings. Most will find more appropriate sleeping spots within a few days or a week.

But if your kitten seems to regard her litter box as her bed, do what I did with Murphy. I persuaded her that a cardboard box stuffed with a fluffy towel is a much nicer napping place. So, when your kitten starts to circle her litter box to get into a good sleeping position, gently lift her out and place her in the cardboard bed. A sleepy kitten is more apt to stay put and rest.

Q Should I be concerned about any health issues when cleaning my kitten's litter box?

A Veterinarians advise owners to follow good sanitary practices when cleaning litter boxes. Wear rubber or plastic gloves when scooping out the deposits. Immediately place deposits into a plastic bag, tie it shut, and take it directly out to your trash can. Then come inside and wash your hands thoroughly in warm soapy water. Clean gloves with antibacterial soap after each use.

There is a slight risk of exposure to toxoplasmosis, a disease caused by parasites harbored in feces. Cats become infected with this parasite by eating raw meat, birds, or mice that carry the disease. Humans also can become infected if they eat raw meat, drink unpasteurized dairy products, or handle infected litter. If your kitten is kept strictly indoors and eats a commercial pet food, the risk of exposure is greatly reduced.

Q I'm pregnant. What precautions should I take with litter boxes?

A During your pregnancy, delegate litter-box cleaning to another household member to completely eliminate your risk of exposure to toxoplasma. The biggest threat of infection occurs during a woman's first trimester of pregnancy. Toxoplasmosis can cause birth defects and mental retardation.

To be on the safe side, always, *always* wash your hands well after playing with or petting your kitten.

Changes in Litter-Box Habits

For weeks your kitty may be a champion of bathroom etiquette, always using her litter box. Then, suddenly, it may seem like she's boycotting the box. Let's look at some possible causes.

Q Why has my kitten stopped using her litter box?

A If your kitten is acting like a little stinker and going outside her litter box, don't unleash your anger. In many cases, a kitten that suddenly stops using the litter box has a very good reason — often related to a medical problem. Your kitten is trying to tell you something is wrong. She may be in physical discomfort or be reacting to a change in her environment.

One thing is certain: the problem will not just disappear by itself, and if left untreated, it can quickly turn into an unpleasant habit.

Your first step in finding out why your cat is urinating and defecating in the house is to book an appointment with your veterinarian to give your kitten a thorough medical exam. Common health problems linked to litter-box avoidance include cystitis (bladder infection), feline lower urinary tract diseases, parasitic worms, food allergies, diabetes, diarrhea, constipation, impacted anal glands, kidney failure, and tumors.

Always rule out a medical cause before you consider something a behavior problem.

Q My kitten seems to be having difficulty urinating. What might be wrong?

A Difficulty urinating can be caused by a host of urinary tract diseases. Don't delay. Bring your kitten to the veterinarian before the situation becomes life threatening. Urine trapped inside the body is a toxin that can poison your kitten.

A kitten with bladder problems will often display the following symptoms. If your kitten is showing these signs, have her examined by a veterinarian.

- Urinates only small dabs in the litter box
- Steps in and out of the litter box and paws at the litter but does not urinate
- Cries when she urinates in the litter box
- Has blood in the urine
- Starts urinating in inappropriate places, such as on your pillow or the rug

Q My veterinarian couldn't find any medical problems, and my kitten still won't use the litter box. Now what?

A If medical problems have been ruled out, your kitten may be sidestepping the litter box because of the odor. Kittens like clean litter boxes and may go elsewhere if the litter box is dirty and smells.

If you dutifully clean out the box daily and change the litter weekly and the behavior persists, try some of these strategies:

- **Change the type of litter and put it in a new, clean box.** Some cats develop allergies to certain litters. There are lots of types on the market, ranging from the clumping kind to wheat litters. Avoid litters that are perfumed, especially those with floral or lemon scents. Cats abhor such odors.
- **Avoid overfilling the pan.** Keep the litter at no more than two to three inches deep.
- **Stop using enclosed litter boxes with lids.** The urine smell gets trapped inside, encouraging some kittens to go elsewhere to avoid the odor.
- **Keep your kitten's litter box far from her food and water bowls.** Kittens won't eliminate near their food source.
- **Sprinkle baking soda into the litter to help reduce its odor.** Baking soda effectively absorbs unpleasant odors.
- **Scrutinize the surfaces.** One of the most common reasons healthy kittens eliminate outside the litter box is because they find some other place or surface more attractive. They may regard a covered litter box as too confining or be frustrated by pan liners. Or, the box may be too close to large appliances that make frightening noises.
- **Check the scents.** Kittens find strong floral room deodorizers and citrus smells highly offensive. Keep these smells at least five feet away from the litter box.

FELINE FACT

Many kittens are particular about the type of litter they'll use, and you may need to experiment to find a litter that your kitten is comfortable with. A scientific study conducted by Peter Borchelt, a certified animal behaviorist in New York City, revealed that most cats seem to give the paws up to fine-grained sandy litters that are not dusty or heavily scented.

The key to solving problem elimination is to make the litter box more attractive while making areas where the kitten is house soiling unattractive, say animal behaviorists. Sometimes just cleaning the litter box more frequently or changing its location and keeping the cat away from soiled areas may do the trick.

Q What clues might my kitten give me to let me know that she doesn't like the current brand of litter?

A Watch for these telltale signs: if your kitten stands on the rim of the litter box or eliminates right beside the box, she is probably trying to avoid contact with the offensive litter.

Q There is no medical problem, and the litter box is super clean, but my kitten is still going elsewhere. Might there be any other explanation?

A One possible trigger: stress. Kittens, just like people, get anxious and upset. When they do, their fight-or-flight response kicks in and they experience an increased urge to urinate and defecate outside the litter box, explains Larry Lachman, cat behaviorist and author of *Cats on the Counter* (St. Martin's Press, 2000).

In addition, if kittens feel their territory is being encroached upon, they will "mark it off" in an attempt to reclaim personal boundaries, either by defecating or spraying urine on vertical surfaces, such as walls and curtains. The key to solving this behavior problem is to target the underlying anxiety in order to prevent the unwanted behavior from becoming a habit.

> **FELINE FACT**
>
> House soiling is a common behavior that can be extremely frustrating for pet owners. In fact, house soiling ranks as the number one reason cited by people who surrender their cats to animal shelters.

Dr. Lachman says that anxiety is the most frequent cause of both house soiling and territorial spray marking. Frequent triggers that can lead to such underlying anxiety include:

- Moving to a new home
- The arrival of a new family pet

- The addition of a new baby to the family
- The departure of a family member or pet due to relocation, death, or other reasons
- Heightened tension and stress in the household due to increased work pressures or family problems (kittens sense the moods of their human companions)

Q If stress is the reason why my kitten is eliminating elsewhere, is there anything I can do?

A If stress is the trigger, Dr. Lachman recommends these tactics:

- Make the target areas where your kitten has urinated or defecated difficult to get to or unpleasant. You can do this by blocking access to the targeted area by shutting the door, laying down aluminum foil in that area (kitties dislike walking on this surface), or spraying the area with citrus scents.
- Engage your kitten in play or give her treats in the areas where she had previous house-soiling accidents so that she learns to regard these places as fun places, not convenient auxiliary bathrooms.
- Clap your hands, blow a whistle, or rattle a can of pennies if you catch your kitten in the act of eliminating in an off-limits location. Don't yell at or hit your kitten.
- Check with your veterinarian to determine whether an anxiety-reducing medication might help.
- Bond with your kitten. Twice a day, take your anxious kitten into a quiet room or area of the home and engage in a quiet, friendly talk. For fifteen to twenty minutes, gently caress her and heap on treats and toys. Make this a special one-on-one time between you and your kitten to build her confidence.

The trouble with a kitten is that . . . eventually, it becomes a cat!

— Ogden Nash

Q Yikes! My kitten is starting to spray the furniture and walls. What can I do?

A Scent marking, or spraying, is one way cats communicate with other cats or people. Intact males tend to leave pungent odor marks more often than neutered males or spayed females.

Marking is a cat's way of telling another cat that it is not happy or feels threatened or challenged. Quite often, the problem arises among cats living in the same household or when an indoor cat sees a stray through a window or door. It's the cat's way of saying, "Hey! This is my place. To prove it, I'm leaving my calling card. So scat."

What can you do? If the spraying occurs due to sibling rivalry between your cats, get them spayed or neutered. Try separating them to let tempers cool down.

If the cause is related to an unwanted four-legged feline trespasser, keep a squirt bottle filled with water near the door and spray it when the cat comes around. Also block the view of the window or door so that your indoor cat can't see the outdoor cat.

If all options fail, check with your veterinarian about giving your cat anti-anxiety drugs. These medications should be used as a temporary solution, in conjuction with a professional behavior-modification program, to correct this problem.

Q We just moved into a new home and my kitten is starting to urinate on the living room carpet. She has always used the litter box before. What can I do?

A Cats are very turf-oriented. If the carpet was there when you moved in and the family had pets, there is a good chance that your kitten is urinating on those spots to mark her territory. In cat language, she is conveying, "Hey! I've got to cover up these old pet smells. I live here now."

First, sanitize the area with a commercial enzymatic cleaner that effectively breaks down and removes protein molecules from the urine, rather than merely masking the odor. Second, dissuade her from repeat visits by covering the area with aluminum foil, a texture cats detest. Third, consider placing her food bowl near the stained area to further discourage her from leaving her mark. Or place lemon or orange peels on the area; cats detest the odor of citrus.

Q My kitty is starting to defecate behind the couch. Why is she doing this — and how I can make her stop?

A First, don't panic and don't scream. Second, get the stool sample over to your veterinary clinic where it can be tested for the possible presence of parasites. If parasites are detected, medications should clear up the problem.

If parasites aren't the problem, it's time to initiate your second-tier strategies, says Roger Valentine, a veterinarian who specializes in cat care in Santa Monica, California. He says that there's a good chance you have a kitten that likes to use an enclosed area as a bathroom.

Here's how to test this theory: Put your cat in a room with nothing that she can get under or behind. Position two litter boxes: one out in the middle of the room and one covered with a cardboard box lid (cut a door opening) positioned on top. See which one your cat chooses. If she picks the second one, that's your clue that your kitten prefers a little privacy, so buy litter boxes with plastic lids. At the same time, booby trap your kitten's behind-the-couch bathroom site by placing double-sided tape there. This is a sticky solution to a stinky problem, and your kitten will avoid it.

Q How can I stop my kitten from using my potted plants as a backup bathroom?

A Think like a kitten. That dark, soft soil makes for a nifty place to leave some deposits. Here's how to fight back: cover the potted soil with aluminum foil or a cardboard cover with watering holes. You can also try sprinkling in mothball crystals. All of these tactics are designed to make it a frustrating challenge for a kitten to use the pot as a potty place.

Other Litter-Box Strategies

Bet you didn't realize there was so much to litter care. Well, guess what? There's more!

Q How can I stop my kitten from inappropriate elimination?

A You need to outfox your kitten by making her want to use her litter box again.

Your first step should be cleaning up the areas where the kitten has urinated. Start with a mixture of white vinegar and water to remove the odor. Dilute one cup of vinegar in one gallon of water and mix well. Allow the surface to dry and then follow with an application of commercial enzymatic cleaner to completely remove the odor. Always test a small, hidden section of the surface you are cleaning — especially carpet — *before* applying enzymatic cleaner to ensure that the chemicals don't cause discoloration.

If the urine has been there awhile, treat the spot several times with vinegar and water. Treating these areas thoroughly will decrease the likelihood of repeat problems.

The best way to curb inappropriate elimination is to intervene as soon as you see the behavior developing. There are a variety of reasons why a kitten would decide to use a corner of your bedroom instead of her litter box for her toiletries. Kittens are very susceptible to stress when household routines get altered. Even the arrival of a new couch can trigger some sensitive kittens to urinate outside the litter box. Your job is to identify the cause behind the bathroom miscues.

You will also need to find out if the kitten is spraying urine to mark territory or if she's just squatting to urinate in a new location. Many kittens begin marking as a result of sexual maturity. Spaying or neutering by six months of age cures most kittens of this desire.

SANITIZING 101

Fighting litter odors often requires a three-step approach:

- Baking soda
- Vinegar and water mixture
- Commercial enzymatic cleaner

Q What's the best way to get rid of cat urine odor in my house?

A Cat urine can be very powerful and pungent. You must first locate the target spots. Conduct a sniff-and-detect search by using your nose and your eyes, says Kit Jenkins, animal behavior manager with the Dumb Friends League, Denver, Colorado, one of the oldest, largest, and most reputable animal shelters in the United States. (Don't be insulted by the shelter's name. Back in 1910 when the shelter was founded, *dumb* used to describe those who could not speak. Organizers adopted the motto, "We speak for those who cannot speak for themselves" — nameless animals.) The nose part is easy. At night, use black light to locate new and old urine stains. Simply turn off the lights, and the soiled areas will magically appear.

Now, it's time to come clean. Avoid using steam cleaners to clean urine odors from carpet or upholstery. The heat from the machines will permanently set the odor and the stain by bonding the protein into any man-made fibers. Also, avoid using cleaning chemicals with strong odors, especially ammonia. All this does is reinforce your kitten's desire to remark that area with urine.

> **FELINE FACT**
>
> Cats can see six times better in the dark than you can.

Instead, clean carpets and upholstery by following these steps:

1. Soak up as much urine as you can with layers of paper towels and newspapers. The more fresh urine you can remove before it dries, the easier it will be to get rid of the odor completely. Place a few paper towels on the urine spot and cover them with a thick section from your newspaper. Stand on this stack for about one minute. Remove and repeat until the area is slightly damp.
2. Take the damp paper towel/newspaper stack to the litter box in full view of your ever-watchful kitten. Let her see you put it temporarily in the litter box and watch your positive body language, which should communicate this message: "Hey, kitty, look here! This is the best place to urinate in the entire house." Be positive, not punitive.

3. Rinse the urine spot with clean, cool water. Blot it up with a clean, absorbent towel or a wet-vacuum.
4. Use a high-quality pet-odor neutralizer on the cleaned area. They are available at pet stores. My favorites are Simple Solutions (Branton, CO) and Nature's Miracle (Pets 'n' People, Inc., Rolling Estates, CA). Both effectively break down the proteins in the urine and make the odor disappear for good. A vinegar-and-water solution works well on tile and hard surfaces.
5. Remove and replace padding under carpet if you realize that a spot has been a frequent target of spraying.

For rugs and other washable items, use your regular detergent and add a one-pound box of baking soda. If necessary, do a second wash, this time with an enzymatic cleaner.

Q What can I do to lure my kitten back to the litter box?

A Start by offering a variety. Try a different style of litter box, one covered and one uncovered. Experiment with different litters, but when making the switch always gradually replace the old litter with the new one over a five-day period. Avoid litter material that contains heavy deodorizers or perfumes.

Practice patience. It may take a week, or it may take six weeks, for your young cat to adjust to the routine use of the litter box again. And always remember that your kitten is not doing this because she is mad at you or wants revenge for your bringing in that high-energy puppy. Your kitten is simply avoiding something that she regards as unacceptable — in this case, the litter box.

Q My kitten seems super-smart and willing to learn. How can I teach her to use the toilet?

A So, you want to say goodbye to cleaning the litter box? Your dream may come true if you practice patience and are blessed to own a willing young cat.

Here are the key steps to a flushing success:

1. Go slow. Some cats learn quickly. Others never do. Be patient.
2. Always keep the door to the bathroom open when it's not occupied.
3. Post reminder notes to household members to keep the toilet lid up and the seat down.
4. Start by locating your cat's litter box next to the toilet. Leave it there for a few days so that your cat gets used to this area.
5. Elevate the litter box two or three inches by placing it on top of a sturdy box.
6. Keep raising the height of the litter box every few days until it is even with the top of the toilet seat.
7. Find a bowl that fits under the seat and will sit inside the bowl of the toilet. Some people use a double-boiler pan, making sure that the lip of the pan rests on both sides of the toilet bowl. A bonus for you: the pan is easy to lift and remove when *you* need to use the toilet. And, the pan is easy to keep clean. Fill this bowl or pan with your cat's regular litter and remove the litter box from the bathroom.
8. Encourage and praise your cat when she uses the pan in the toilet. Recognize that it may take your cat some time to feel comfortable with this new "litter box." Speak in a soft, reassuring voice.
9. Watch the position of your cat's feet. The more paws she is willing to rest on the toilet bowl's outer rim, the better chance of success you will have.
10. Begin to remove more and more litter from the toilet pan as you notice that your cat is mastering this new toilet technique. Work down until there is only about one-half cup of litter left.
11. Start adding water to the toilet pan, beginning with a few tablespoons and working up. This helps your cat get used to water without any fear.

Cats seem to go on the principle that it never does any harm to ask for what you want.

— Joseph Wood Krutch

12. Remove the toilet pan once you've filled it halfway with water. It's time for your cat to use the toilet free of any training tools.
13. Save the flushing until after your cat has left the bathroom. You don't want to startle it with the sound of the swirling water.

Kitty Caper: Bathtub Greeter

I rescued Pork Chop from a dump near my veterinary hospital. He was young, skinny, and scared. Once he got used to prowling the premises and feeling at home in my house, he possessed incredible timing. As soon as I would step out of the bathtub, he would race over and rub against my wet leg. Yuck! His wet fur was matted against my leg. He also liked to walk between my legs while I was cooking in the kitchen.

I needed to draw some boundary lines for Pork Chop to keep my legs from getting hairy in the bathroom and from keeping a boiling pot of water from spilling on him in case he accidentally tripped me in the kitchen.

My solution? I started to blow air into his face and then say, "Uh-uh." Blown air to a cat is like you biting into a sour lemon. Cats hate air in their faces. It took a few days, but Pork Chop got the message. Now, he waits outside the bathroom door and watches me cook from a chair a safe distance away.

—Soraya Juarbe-Diaz, D.V.M.
Southern Florida

CHAPTER SIX

Adopting a Kitten

You might not like what they have to say, but cats will never deceive you.

— Lewis Carroll

There is certainly no shortage of kittens on this planet. In fact, there are far too few homes for the number of kittens born every year.

If you're planning to adopt a kitten, remember this is a baby in fur that needs your attention, your guidance, your love. In return, he can grow up into a playful, affectionate companion that thinks you're the cat's meow!

Choosing the Right Kitten

So many choices, so many decisions. Which kitten is right for you and your household? It isn't always an easy answer but, to help you narrow your choices, let's review a few things before you head out to a shelter or to a professional breeder.

Q What type of kitten will best suit my household?

A Before you make a decision, have household members make a list of what they want and don't want in a kitten. It's important to make individual lists without any influence from other family members.

- Am I allowed to have a pet where I live?
- Is anyone in the household allergic to cat dander?
- Do I like short-haired or long-haired cats?
- Do I want a male or a female?
- Do I want a lap lounger or a playful pal?
- Am I willing to care for a full-of-energy kitten or do I prefer a calmer, older cat?
- Is my household quiet or noisy?
- Do I spend a lot of time at home or do I travel a lot?
- Am I willing to clean the litter box, groom the kitty, provide it regular medical care?
- Do I want two kittens instead of one?

Then meet as a group to review your findings, being sure to discuss and resolve any conflicting answers on the lists. Making these lists help people make better adoptions.

Q From whom should I adopt my kitten? The local shelter, a breeder, my neighbor, or a pet supply store?

A Let the kitten buyer beware! Where you get your kitten can play a powerful role in his health and temperament. You shouldn't adopt the first fuzzy face you see simply because some stranger standing outside your supermarket is handing out free kitties.

"Free to a good home" — beware of that phrase when kitty shopping. You think you've got a bargain, but in life there are no free kittens. Resist the temptation to adopt a kitten from someone you don't know who is handing them out for free. There is a good possibility that the mother cat was never tested for contagious diseases, such as feline leukemia, which may have been passed on to her kittens. If you bring home such a kitten as a companion for your cat, the kitten may pass on this incurable disease.

Let's look at safe sources for kitten selections:

- **Your local animal shelter.** This is my number one choice because there are so many homeless kittens and cats in need of good homes. More and more shelters are spending time socializing their animals, which increases their chances of wowing you when you walk by their cages. It also improves the number of good adoptions. Reputable shelters will spay or neuter these kitties before they are adopted and will have them thoroughly checked by a veterinarian who will give them the necessary vaccinations. Don't be offended if the shelter also does a thorough background check on you. Yes, the shelter will call your landlord to make sure pets are okay. It's a good measure that prevents a swift return to the shelter for a much-wanted kitten in a no-pets apartment complex. Shelters also sometimes choose not to give kittens to people who travel excessively or who appear unready for a pet.

> **FELINE FACT**
>
> Newborn kittens snooze nearly round-the-clock, spending 90 percent of their time asleep. By age four weeks, kittens spend 60 percent of their time napping.

- **A pet supply store.** Stick with stores that provide health guarantees and ones that aren't in the "kitty mill" business. More and more national pet supply store chains, such as PetsMart and PetCo, are offering space in their stores to local rescue groups or shelters that have veterinarian-examined kittens. I like this concept. The shelters have a great place to showcase their homeless pets and the stores are helping to reduce pet overpopulation and the need to euthanize.
- **A reputable breeder.** If you want a purebred kitty with papers to prove it, then I recommend dealing with an experienced breeder willing to answer your questions and willing to let you visit his or her premises. On page 104 you'll find a checklist that will help you select a good breeder. Or, you can adopt a purebred from a purebred rescue organization.

Q How can I tell if the kitten I'm adopting is healthy?

A The final verdict will come from a veterinarian after he or she gives your new kitten a thorough physical examination. But, when deciding which kitty to adopt, here are some healthy signs to consider:

- Eyes should be bright and clear.
- Nose should be clear and free of discharge.
- Ears should be clear and free of dirt or odor.
- Mouth should feature pink gums and clean teeth, and have no signs of sores or ulcers.
- Coat should be smooth, clean, and free of fleas, dandruff, mats, or cuts.
- Anal area should not have dried waste or be discolored.
- Body should be lean without a protruding belly.

Q How can I learn more about a specific breed of cat?

A Tap into the Internet. I recommend that you check out the reputable and helpful Web sites for The International Cat Association and Cat Fanciers Association, listed on page 194. Both are loaded with oodles of information on each recognized breed and graciously link you to other great cat care sites.

Q In selecting a breeder, what types of questions should I ask to make sure I'm dealing with a reputable professional and not a backyard breeder?

A Be diligent in the search for your perfect kitten. And be selective. After all, you're shopping for a feline friend that may be with you for at least ten years, possibly even longer.

Jot down these questions and ask them when you speak with a breeder:

- **How old are the kittens when they are ready to be adopted?** Steer clear of breeders anxious to sell kittens younger than twelve weeks of age. Kittens need ample time to socialize with their littermates, be weaned from their mothers, and reach some early physical and cognitive developmental milestones.
- **What's the health status of this kitten?** A good breeder can tell you about the family history, including any genetic problems, and provide records of vaccinations and physical examinations done on the kitten.

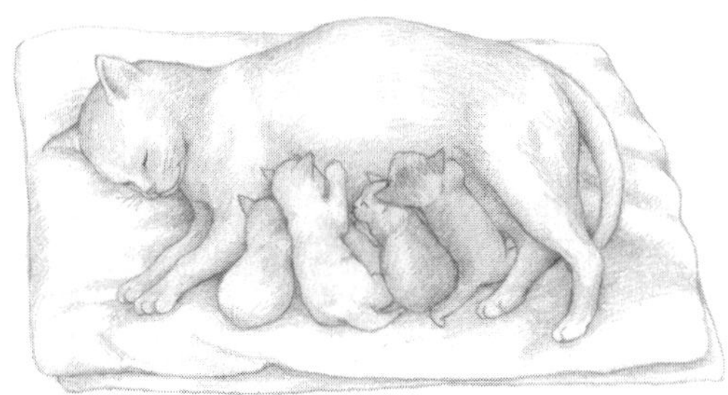

- **What have you done to socialize this kitten?** Good breeders recognize the power of touch on growing kittens. They also make sure the kittens learn to use the litter box and scratching post. You should see lots of toys around, too. You should also be able to visit with the kitten's mom to get a sense of her health and temperament.
- **Is this litter registered, and have you shown the parents in cat shows?** Often breeders active in cat shows are concerned about improving the breed. And, they will be willing to provide the registration papers on the litter.
- **Where does the kitten live?** Good breeders raise kittens in their households, not in cages in boarding areas adjacent to their homes.
- **What is your cattery's background?** Stick with breeders who voluntarily participate in the Cat Fanciers' Association inspection program or those who willingly refer you to the veterinarian who has cared for the litter. Find out how often cats are bred. Good breeders do not breed more than one litter per cat per year.
- **Do you offer a written health guarantee?** Responsible breeders encourage you to have the prospective adoptee examined by your veterinarian and are willing to offer a refund or accept the return of the kitten if the veterinarian discovers a health problem.

Make Mine a Pedigree, Please

Don't like surprises? Interested in possibly entering your fancy feline in cat shows? Then a pedigree may be your best bet. By selecting a purebred, you can often match a kitten with your lifestyle. But be prepared to open your checkbook. Pet-quality purebred kittens, on average, come with price tag of $400 — much more if you want a show-quality pedigree. There are about forty breeds recognized by Cat Fanciers' Association and The International Cat Association, two of the largest registries. To help you decide, here's a baker's dozen of the most popular breeds:

Most Popular Breeds

Breed/Pounds*	Looks	Personality
Abyssinian 5–10	Large ears; triangle-shaped face; almond-shaped eyes; short, shimmering coat. Colors are ruddy red, blue, and fawn.	Super-smart and a born jock that craves a packed schedule; demands attention but will charm and amuse you.
American Bobtail 7–20	Wedge-shaped head; high cheekbones; almond-shaped eyes; brawny, athletic body. Spotted or ticked brown tabby pattern.	Bubbly; loves people; up for new adventures and tricks.
Bengal 8–12	Lean, muscular body; triangle-shaped face; golden eyes; glistening shorthaired coats with spots and swirls.	Smart; loves to stalk prey; extremely curious; downright affectionate and friendly. Play with this breed and you'll have a pal forever.
Burmese 7–12	Large, innoncent-looking eyes; compact, sturdy body. Colors are sable, champagne, blue, and platinum.	Definitely a people pleaser; often described as a dog in a cat's body because of his desire to be your constant, loyal companion.
Korat 7–13	Heart-shaped face; luminous green eyes. Silver-blue shorthaired coat.	Super cuddler; prefers a quiet household to a rowdy one; more apt to join you in yoga than to let children dress him up in doll's clothes.

Maine Coon 7–18	Tufts of fur accent paws and tips; big and ruggedly handsome. Colors are blue, red, cream, calico, tortie (multicolor), silver, smoke, and shaded.	Highly adaptable, playful, loyal, easily trained; a gentle giant among cats.
Manx 6–12	Tailless with a thick, dense coat; either shorthair or longhair. Many colors and patterns.	Playful, affectionate; tends to be a one-person cat, but tolerates others; generally gentle and quiet.
Persian 6–11	Short nose; broad face, round-shaped body; short, stocky legs; short tail. Many colors and patterns.	Librarian-like; likes quiet households, perching from a high vantage point, and a predictable routine; shy but loving.
Ragdoll 6–12	Silky, semi-long coat; long, plumed tail; blue eyes; large, muscular body. Colors are seal, chocolate, blue, and lilac; many patterns.	The ultimate lap cat and shoulder rider; prefers quiet homes without rambunctious children or dogs; revels in being pampered.
Russian Blue 8–12	Emerald green eyes; long legs. Silver-tipped blue shorthaired coat.	The ballerina of cats; extremely graceful and agile; die-hard loyal but gets along nicely with children and dogs in the home.
Scottish Fold 7–10	Ears folded forward and down; large, round eyes; shorthair and longhair. Many colors and patterns.	Mellow; charming; always flashing a sweet expression.
Siamese 5–8	Triangle-shaped face; lanky legs; almond-shaped blue eyes. Point colors are seal, blue, chocolate, and lilac.**	Super socializer; chatty cat that often demands attention; loves to learn tricks.
Sphynx 5–8	Hairless; large eyes; pixie face; sturdy body with barrel-shaped chest and thick, rounded abdomen.	Surprisingly active; always affectionate, especially if you offer a warm blanket; very intelligent; accustomed to being stared at for its unusual looks.

* Males tend to be larger than females.

** ***Point colors*** are accent colors found on the tip of the ears, around the face, and on the feet.

Q Pedigrees can be pricey, and I would like to adopt a kitten from my local shelter. Is there any way to determine what a mixed breed's personality and temperament are like?

A Selecting a kitten that matches your lifestyle can be challenging. If you're not getting a purebred, how are you going to know whether your selection will be a lap lounger, a shy cat, or an adventure seeker?

Try looking at the shape of the kitten's face for clues to their possible personalities. Feline faces fall into one of three physical shapes: square, round, or triangle. At least, that's the theory proposed by Kit Jenkins, an animal behavior manager at the Dumb Friends League in Denver, Colorado, after twenty years of studying cats in shelters and the showring.

"A dog's behavior is much easier to predict than a cat's because dog's have been bred for specific job purposes, such as herding and hunting," says Jenkins. "Cats have not been bred for specific jobs. I've found that you can predict their personalities to some degree based on their looks and body types."

Yes, genetics and life experiences play major roles in how a cat thinks and acts, but personality is also influenced by a cat's physical shape, suggests Jenkins.

Here's how Jenkins sizes up cats:

- **Square-shaped cats:** These cats tend to be big and solid with square faces, bodies, and legs. Think Maine Coon or striped tabbies. Jenkins dubs them the "retrievers of the cat world." Eager to please, square cats are affectionate and love to snuggle and give head-butts.
- **Round-shaped cats:** These cats sport flat faces, big eyes, and round-shaped heads and bodies. Think Persian. They tend to be low-energy, easily frightened, submissive cats that gently display their affection to trusted family members.

- **Triangle-shaped cats:** Look for sleek, long, lanky cats with big ears and faces that narrow at the nose and you've met what Jenkins calls "the herding dogs of the cat world." Think Siamese or Abyssinian. Triangle cats are busy, curious, smart, athletic, and exceedingly vocal. They thrive in active households.

Roger Valentine, a veterinarian in Santa Monica, California, who devotes much of his practice to caring for cats, says this "cat geometry" notion can't be proven scientifically but calls it an "interesting concept." At the very least, he says, the theory sparks in-depth conversations about cats.

"I just realized that I have one of each type in my household," says Dr. Valentine. "Scooter is my round-shaped cat who is scared of strangers. Xochi is my square-shaped tabby who is doglike friendly, and Spider is part-Siamese, very vocal, and very athletic."

As for me, only after I applied the "cat geometry theory" did I realize that I own one of each personality shapes: a round, a square, and a triangle.

> **FELINE FACT**
>
> Allergic to cats? Blame your reaction on a cat's dander and dried saliva, not on her coat. Flakes of skin, not hair, produce dander. While you may lessen your symptoms by getting a breed with no hair such as the Sphynx, it's no guarantee that you'll be sneeze-free.

Q I want to get a kitten as a pal for my cat. Any advice?

A They don't like to advertise this much, but most cats are social creatures that enjoy constant companionship. Getting your adult cat a furry playmate can often ease her feelings of boredom and anxiety while you're away from home.

Ideally, you should aim for complementary personalities whenever feasible. If your cat is an extrovert and bold, pair her up with a kitten that appears to be easygoing and willing to play follow the leader.

Be realistic. Most kittens receive a hostile reception from the resident cat. They need time to sort out who is top cat under your roof.

Q **I hear that cats sometimes adopt their owners. Is that true?**

A Each one of my three cats "adopted" me. All three were homeless kittens before we found each other. Little Guy came begging at my screen door at my southern Florida home. I fed him; he winked at me and let loose a full-throttle purr. Callie was separated from her littermates at three weeks of age and was dodging car tires on a busy Miami highway when I found her. Murphy popped out of the bushes near my southern California home and heeled like a dog by my side when I took my daily afternoon walk.

Before I brought each kitten into my home, I took them directly to my veterinarian, who performed complete physical examinations to make sure that they were disease-free.

FELINE FACT

The first cat show in the United States was held in May 1895 at Madison Square Garden in New York City.

Kitty IDs

Once you choose your kitten, it's vitally important that you provide her with appropriate identification.

Q **I plan to keep my kitten indoors. Does she need a collar and tags?**

A An identification tag is a lost kitty's ticket home. Even though you pride yourself on keeping your kitten indoors,

COLLAR SAFETY

In addition to a collar being sized appropriately for your kitten, it also must be of the right type. Snag-resistant breakaway collars and safety collars are best. If your mischievous kitten should get his collar caught while exploring, the collar releases or stretches, respectively, to set him free, minimizing the risk of choking.

you still need to practice some precautions, just in case your feline friend suddenly finds herself outside. Try as you may, you can't always prepare for that back door left partially opened by a visiting friend, a loose window screen, or your frightened kitty wiggling out of your arms as you step out of your car from a veterinary visit.

It's the most sinking feeling to know that your pampered indoor cat is alone, frightened, confused, and hungry somewhere outdoors. Only 2 in every 100 cats plucked from the streets that are brought to shelters are safely returned to their owners, according to recent statistics compiled by the Humane Society of the United States. The prime reason: most cats found are not wearing tags.

Favorite Names

Can't decide what to name your new kitten? Here's a list of the Top Ten Most Popular Pet Names, in order, based on a survey of thousands of pet owners conducted by the American Society for the Prevention of Cruelty to Animals:

1. Max
2. Sam
3. Lady
4. Bear
5. Smokey
6. Shadow
7. Kitty
8. Molly
9. Buddy
10. Brandy

That's why it is important to keep a collar with identification tags on your kitty at all times (except for grooming and bath time). Make sure that the collar fits snugly, but that it's not too tight or too loose. If you can slide one or two of your fingers under the collar easily, that's a good fit. Remember that your kitten keeps growing, so check the collar's fit periodically. You may need to go up a size.

Q Where can I get an ID tag?

A Getting an ID tag these days is as easy as buying cat food. That's because many pet supply stores house do-it-yourself ID-tag-making machines that are easy to operate, quick, and inexpensive to use. In a few minutes, you can put the pertinent info on a tag in the color, size, and shape of your choice. So if your cat does happen to slip out, someone will have a way of contacting you. Veterinary clinics and pet mail-order catalogs also offer a variety of ID tags.

Introductions

Now comes the critical time: you're ready to introduce your new kitten to the other critters in your home.

Q How should I introduce my kitten to his new home?

A Imagine being no bigger than a softball and no taller than a soda can and suddenly finding yourself in this *gigantic* place with *gigantic* people. You may think your home is cozy and comfy, but to a small kitten, your home may be as big — and as scary — as a continent. From a kitten's vantage point, there are too many doors, too many feet, too many corners, too many objects.

Information overload.

Remember the movie *Honey, I Shrunk the Kids?* Well, keep that concept in mind and temporarily shrink your kitten's home environment. On day 1, limit your kitten to one room in your house, preferably the bathroom or a den — a small room without a lot of hiding spots. Be sure to provide toys, a bed, litter box, water, food, and a scratching post.

With you in the room, let this tentative kitten safely explore its new surroundings. Speak in a calm, reassuring tone. Kittens often do what's known as the "perimeter prowl," a low-slinking movement around the walls. When they gain a little confidence and their bearings, they will then explore the interior of the room. Your kitten may also rub his cheekbone against a wall or piece of furniture. That's a good sign. Your kitten is leaving his scent marks.

Within a day or two, your kitten will proclaim this room as his own and may view it as a "safe place" to scurry to whenever he feels a bit uneasy or unsure or simply wants to be left alone to take an uninterrupted snooze.

Gradually introduce your kitten to other rooms in your home. Be sure to keep closet doors shut and block off small, tight places that your kitten may try to wiggle into and get stuck.

Q What's the best way to introduce a new kitty to a household with children?

A Young children, especially those under age seven, may not understand how to treat a kitten without unintentionally hurting it. Children that age aren't physically coordinated and may play too rough by trying to pull the kitty's tail or ears.

The success depends on supervising the kitten-child interaction and in setting up ground rules before you bring a kitten into your home.

Teach your children these do's and don'ts:

- *Do* pet the kitty slowly and softly.
- *Don't* pull on the kitty's ears or tail.
- *Don't* hit, kick, or throw objects at the kitty.
- *Don't* disturb the kitty when he is eating, sleeping, or using the litter box.
- *Do* wash your hands after touching the kitty

And here are some do's and don'ts for parents supervising interactions between young children and kittens:

- *Do* teach your child to treat the kitten — and all animals — with respect. Help your child learn about the benefits of having animals as companions.
- *Do* show your child the right way to stroke a kitty's fur.
- *Do* praise your child when he or she does something nice for the kitty.
- *Do* keep your kitty's nails trimmed so that a claw swipe doesn't turn into a trip to the doctor's office.
- *Do* encourage your child to take part in the kitty-care chores such as feeding and cleaning his litter box.
- *Don't* leave your child and kitty in a room unsupervised.

Q What's the best way to introduce my new kitten to the adult cat in my home?

A This is no time for a direct introduction. In fact, you need to be a bit sneaky at first by actually smuggling in your new kitten.

Making the proper introduction is vital to ensuring a lifelong friendship. Murphy is my newest arrival. Found homeless on the streets of southern California at about six months of age, she became fast friends with my adult cats, Little Guy and Callie, because I heeded these instructions from my animal behaviorist friends. I waited three days before letting them meet.

- **Bribe your cat.** Buy your cat a new scratching post a few days ahead of the new pet's arrival. Okay, so it's a bribe, but your cat will associate this prize with an impending positive change.
- **Be patient.** Solid friendships sometimes take time. Some cats take weeks, even months to become paw pals. Expect to hear a few hisses.
- **Plan ahead.** Select a large bathroom or spare room to house the new arrival. Place cat necessities in that room: food and water bowls, bedding, litter box, and toys, and shut the door so your house cat won't snoop around.
- **Be stealthy.** Bring the new pet in quietly and incognito. Try not to let your cat see you entering the door with this animal to avoid any resentment. Don't pussyfoot around. Walk straight to the new cat room and place him inside and shut the door.
- **Encourage the sniff test.** By now, your house cat will suspect something is different and will be drawn to the door. Let both meet each other by sniffing one another from under the doorframe. This helps them get to know one another on their own terms.
- **Share scents.** After a day or so, take a slightly damp towel and rub it on your new kitten's back. Then rub this towel on your adult cat's back. Take a second damp towel and rub it first on your adult cat's back and then on the new arrival's back. Intermingling scents encourage familiarity.
- **Don't play favorites.** Spend quality one-on-one time with each cat. Pamper them with plenty of praise, hugs, and treats. Make each one feel special.
- **Switch rooms.** After two or three days, switch places. Put your house cat in the spare room for a few hours and let the

new arrival check out the rest of the house. This helps prevent any possible turf tussles.

- **Make the introductions.** You're finally ready for the face-to-face introduction. Let your house cat be free to approach the new arrival that you place inside a carrier or on a leash. Let them have plenty of time to approach and sniff. Expect a few hisses — it's your adult cat's way of declaring, "Hey, I'm the boss around here."
- **Encourage interaction.** Gradually increase the exposure time of the animals to each other. Give them both food treats, always offering a treat to your adult cat first.

Once you feel confident that the two felines can get along, leave them alone unsupervised.

Q What's the best way to introduce my kitten to my dog?

A You need to think like a dog — and a kitten. Understand what motivates each of these species. And, like you do with your children, you need to establish consistent household rules for both of your pets.

Dogs are pack animals that look to you, the leader, for guidance, direction, and approval. Cats tend to be territorial. If they feel that their home turf has been disturbed or threatened by a dog, they may feel anxious or defensive.

Before making the introductions, locate the kitten's food and water bowls and litter box in an elevated place, out of dog's reach. Provide safe perches and rooms for your kitten that will be off-limits to your dog.

When you bring the kitten home, keep him in a separate room from your dog for a few days — out of sight but not out of scent from one another. Then switch rooms so that they can investigate each other's scents.

For the first face-to-face meeting, keep the dog on a leash or inside a crated kennel. If your kitten flees the room, don't give chase. Your dog may misinterpret your actions as signs of support and be motivated even more to give chase. If your dog starts to lunge, step on her leash and say, "Leave it!" — a command that, ideally, she knows and can perform before you make the kitty introduction.

Try to schedule feeding times in the same room at the same time, if possible. If your dog and kitten eat at the same time, that develops a powerful connection for them. Just make sure that the kitten's food is elevated in a place where he feels safe, so he can eat without any interference from your dog.

Multiple Pets

It is possible to achieve harmony in a multipet household. In many respects, you must act as mediator. Let's look at how to reduce fur-flying episodes between your new kitten and cats and dogs you already own.

Q How can I discourage "sibling" rivalry between my cat and my new kitten?

A Make sure that the kitten has his own food, food bowl, and water bowl. And, as adorable as that kitty may be, you really need to increase the time you spend lavishing attention on your resident cat. Your cat needs to feel that she rates supreme, not slighted.

When you offer treats, always serve your cat first and the kitten second. The same goes for introducing new toys or playtime. In subtle ways, you are trying to show your cat that this kitten enhances your household and your cat's life.

In some cases, the best that you can expect is a peaceful coexistence, a tolerance of one another. That's okay. But, with a bit of luck, your cat and kitten will bond over time, start to play together, groom each other, and sleep near each other. This scenario is more likely to happen if you let them work things out between themselves and don't force them together.

Q Why does my dog insist on chasing my kitten?

A Dogs are descendents of wolves and have a natural instinct to go after anything that darts and dashes — like your high-energized kitty.

This desire to chase, catch, shake, and kill varies in intensity among breeds and individual dogs. There are exceptions, but generally speaking, beagles, greyhounds, and terriers have been trained for the bloodthirsty stalking and killing of speedy prey. Less violent Irish Setters have reputations as "soft-mouthed" predators because they retrieve killed prey for hunters without leaving any teeth marks.

On farms and prairies, shelties, sheep dogs, and other herding breeds have been raised to guide sheep and cattle. Without sheep or cattle around, this breed may try to herd cats along specific paths. Bullmastiffs have been bred to be territorial, so they usually stop chasing once the trespassing cat crosses the property line. The least chase-minded are toy breeds, such as Chihuahuas and Pomeranians. They'd rather secure a spot on your lap than expend energy dogging a cat.

Dogs often chase cats because they want to socialize, to get to know each other. The problem is that dogs and cats don't speak the same language. Dogs can't tell cats of the playful but benign intentions behind their coming in closer. So, the canine message of "Let's play" may be misinterpreted by kittens as "You're prey."

Pursued kittens often worsen these situations. When a cat knows he's the target of a dog's glare, the cat instinctively thinks, "I need to escape. Now," and flees for safety. The dog senses the cat's fear, triggering her prey drive and sparking the chase.

Q How can I teach my dog to stop chasing my kitten?

A Outfox your dog by redirecting her chasing tendencies. But before you can teach your dog not to chase cats, you need to make sure she obeys basic commands such as "Sit," "Stay," and "Leave it." Every day, reinforce these basic commands with your dog, on and off leash. Each time the dog obeys, give her praise and perhaps a food treat.

Then look for pre-chase signs. A dog doesn't try to disguise her desire to chase a kitten, and it's your job to be able to spot these signs and stop the chase before it starts. Look for the biggest giveaway: the dog's cold, hard stare at the kitten. Also, when a dog is about to chase, she hunches forward and the hair along her spine sticks up.

If your dog starts to chase, call her by name followed by the command "Leave it." Use a stern, low-pitched tone. Do not yell, scream, or use high-pitched cries, which tend to increase the panic of the situation rather than calm it.

Follow your "Leave it" command with a noise that will break the stare of your dog on your kitten. Clap your hands, slam a book shut, or shake a can filled with coins.

As soon as your dog makes eye contact with you, redirect her attention to chasing a favorite toy in the opposite direction of the fleeing feline. Spend time playing with your dog so that she can expend this energy and praise her for playing with the toy.

Q How can I tell the difference between a playful romp and a deadly pursuit by my dog and kitten?

A Some kittens like to play hard to get. They relish the playful chase — without growls or hisses. A kitty that feels frisky will often thump his dog pal on the head with a clawless paw to start the friendly chase. It is the feline's way of saying to the dog, "Tag, you're it." The kitten's posture should be relaxed. The dog will often respond to the thump with a friendly play bow, front paws extended out and head tucked into the chin. Her tail is wagging and her ears are relaxed. During the chase, the kitten moves silently without a single hiss or shriek.

Happy kittens play silently, whereas fear-filled ones or those that have had enough, release a cry, hiss, or spit. Be aware that play-minded dogs let loose high-pitched yelps during the chase, whereas prey-driven dogs deliver low growls.

The minute you hear a hiss or a low growl, intervene, sternly say "Stop," and end playtime. Don't worry if you made a mistake. If your kitten truly wants to keep playing, he will. In time, your kitty may compromise and agree to play with your dog, but on feline terms. And, your dog will learn to respect your kitty.

Kitty Caper: Meet Maggie, the Dog-Testing Cat

People in the Pasadena, California, area shopping for a dog to add to their cat household often seek the advice of a wily feline named Maggie. A brown-striped tabby, Maggie has one job in life — official dog tester — and she performs it flawlessly at the Pasadena Humane Society. For more than nine years, she has served as protector of household cats, bravely standing up to shelter dogs being considered for adoption in homes with cats. Dogs that pass the Maggie test — that is, don't growl, lunge, or become dominant — win adoption.

Good canine candidates for cat households yield to Maggie by avoiding direct eye contact, backpedaling away from her, or trying to hide behind the leash-carrier's legs when introduced to her inside the animal shelter's volunteer center office.

The size of the dog doesn't matter to the otherwise mellow Maggie. She's tried to pounce on Dachshunds and never surrendered an inch to highly aggressive Dobermans.

Unfortunately, Maggie's talents are limited to dogs. She has no patience for cats. "Maggie goes out of her way to spit, hiss, and yowl at the other cats housed at our shelter," says Liz Baronowski. "She's no cat's cat."

—Liz Baronowski, Director of Humane Education
Pasadena, Calif.

On the Go with Kitty

We live in a mobile society. During your kitten's lifetime, she will definitely ride inside a car and possibly inside an airplane. She will probably move with you to a new location. You can make a move, or even a trip to the veterinary clinic, far less daunting for your kitten by conveying a sense of excitement and adventure and speaking to her in upbeat, reassuring tones. And you can give yourself peace of mind by always making sure your kitten wears an identification collar, just in case she gets lost.

Traveling by Car

You may like how your kitten purrs, but your kitten may not. The engine sounds, the rapid movement, and the enclosed space can make your kitten feel uneasy during a car ride. Let's look at ways to acclimate your kitten to the car.

Q I'm adopting a kitten, and I need to bring her home in my car. Is it okay to have someone hold her in the backseat?

A Even small kittens can be master shredders when spooked. To protect both your passenger and your first-time-in-a-car kitten, place the kitten inside a pet carrier in the backseat. Your passenger can ride in the backseat and offer some verbal reassurance to your kitten while you drive.

Q My kitten has never been in a car. What's the best way to ride with my kitty?

A *Never* let your kitty lounge on your lap while you're driving. All it takes is one sudden stop and your kitty becomes a furry projectile inside your vehicle. Instead, get your car cat a crate. Check your pet supply store for carriers that feature a quick-release adjustable snap that attaches to the car's seatbelt.

Introduce your kitty to the crate inside your home. Leave the door open and entice her to explore it by tucking in a few food treats or her prized toy mouse. Decorate the crate like a mini-kitty condo with plush blankets or a couple of your old, unwashed T-shirts or sweatshirts.

Once your kitten identifies the crate as a cool source of treats and toys, you're ready to put her inside it inside your parked car. Spend five minutes or so and let her get used to this new setting. Be sure to run your air-conditioner if you do this during the summer or car heater if it's during the winter. *Never* leave your kitty in the car alone. Speak in a calm, soothing voice. Bring a few tasty treats or one of her favorite toys. Try to do this a few times a week.

If you don't have a garage, put your kitten on a leash — or better yet, a harness — inside her carrier and take her to your parked car. Once she is safely inside the car and the doors are all shut, let her out to explore while the engine is off.

Now, you're ready to make a short trip, say a mile or two. Buckle the carrier into the seatbelt, and be on your way. When you return home, park the car, and give your kitten a food treat before you remove the crate from the car. This way she says to herself, "Hmmm. This isn't so bad. When the engine shuts off, I get served a tasty treat. Bring it on."

You're conditioning your kitten to associate the car with pleasant experiences, not that once-a-year terrifying trek to the veterinarian. Kittens may be young, but they're savvy. And, they are willing to play the odds. If nine out of ten times they go into the car and get treats, they'll tolerate the occasional visit to the vet.

Q What should I pack for my kitten on a long road trip?

A For long car trips, create a mini-litter box by using a cardboard shoebox. It fits nicely inside the crate. Bring a couple favorite toys; two sets of bedding (in case of a kitty potty accident or vomiting due to motion sickness); dry food stored inside waterproof containers; spring water in resealable, plastic containers; and a small, nonspill drinking bowl. And, don't forget to pack a pet first-aid kit.

To prevent your kitty from overheating or becoming dehydrated, keep a spray bottle of water within reach. On long, hot rides, spritz your cat's face and paws to cool her down. Use the air conditioning during the day or try to drive during the cooler times of day or during the night.

Never leave your kitty inside a parked automobile even for a few minutes, especially on a hot day. The inside of a car can quickly heats up like an oven, reaching 120°F and higher.

Also, your kitty should be up-to-date on all her vaccinations as well as monthly flea and tick medications *before* heading out on a long trip. You don't want to be itchy and scratching on the interstate.

Part Dog?

If your kitten loves riding in the car with you, you're part of a small but mighty minority. According to the Travel Industry Association of America, only 7 percent of people regularly take their cats inside their cars to work, on vacation, and to visit friends.

Vacation Time

Finally, you have a chance to take a break from the office and go on vacation. You are looking forward to relaxing and having fun. Would your kitten be happy to join you or would she prefer to stay at home? Give this careful thought before dashing off for your great escape.

Q Should I bring my kitten with me on vacation?

A Before you envision you and your kitty lounging poolside in a Caribbean resort, think about how much she would enjoy traveling. Most felines crave routine and are homebodies. As much as you may miss her, it may be better for her to stay behind and be the keeper of your home.

Some kittens can get so stressed out by all the strange sights, sounds, and smells of travel that they begin to vomit, pant, drool, and have diarrhea.

You also need to contact the hotels you intend to stay in to make sure that pets are allowed. Sure, it's easy to sneak in a kitty, but why take that risk? Consult travel guides offered by AAA and Mobil for a listing of pet-welcoming hotels.

> **FELINE FACT**
>
> Sixteen percent of cat owners polled in a national survey reported that they take their feline friend with them on vacation. Another 23 percent bring their cats on visits to schools or health care facilities.

Q We're planning to take a two-week vacation this summer. How can I locate a pet sitter in my area?

A Cats are much more independent than dogs, but if you plan to be away from home more than a couple days, you should rely on a trusted neighbor or a professional pet sitter to cater to your kitten's needs.

If you're new to town or don't know how to reach a professional, licensed pet sitter, contact the National Association of Professional Pet Sitters (see page 195).

Photo IDs

Admit it — you love taking photos of your kitten. I have a drawer full of photos of my cats sleeping, playing, and giving me those adorable over-the-shoulder glances. Take a current photo of your kitten and mount it on your refrigerator door with a magnet. That way if your kitten ever becomes lost, you can quickly grab this photo and show it to neighbors.

You can also locate great pet sitters by asking for recommendations from your veterinarian or pet-owning friends.

For peace of mind, make sure the pet sitter provides you with references, is bonded, insured, and has experience.

Arrange to have the pet sitter visit you and your kitty at your home a few days before your vacation starts.

Q What instructions should I give the pet sitter?

A A professional pet sitter can be your home-alone kitty's best pal. He or she can keep your cat well fed and not starving for attention.

Here's a handy checklist to run through with a pet sitter during your face-to-face meeting.

- **Provide your kitten's name and favorite nicknames.** My Little Guy will also answer to Dude. It lets the kitty know that this "stranger" is okay.
- **Divulge your kitty's quirky habits.** Callie, my calico, loves to chew on cardboard boxes. Murphy loves coming when I whistle the theme from *Jeopardy!* Some kitties will dart under the bedspread when the doorbell rings; others will rush to greet visitors at the door.
- **Inform how much, what, and when you feed your feline.** To make sure your kitty is eating while you're away, advise the pet sitter to pre-measure each meal each day.

- **Leave contact information.** Provide the sitter with contact information for your veterinarian, a close relative, or neighbor in case of emergency and you're not available.
- **Go over the particulars.** Offer instructions on how you clean the litter box and how you dispose of the litter.
- **Make some introductions.** Introduce your pet sitter to a few of your close neighbors so that they won't be suspicious when the sitter enters your home.
- **Make a list.** Post all of this information on your refrigerator door so it is easily accessible to the sitter.

Many pet sitters are also willing to pick up your newspapers and mail, and water the plants. Some agree to actually housesit — stay in your home while you're gone, which gives your attention-seeking kitty a captured audience.

Q What other options do I have besides a pet sitter?

A Trusted friends, neighbors, and relatives can temporarily take on the kitty care duties. The key here is trust. You want someone who you can depend on to visit your kitty every day while you're gone.

Or, you can take your kitty to a boarding facility at your veterinary clinic or kennel. These are best for easy-going, well-adjusted kittens. They can handle the new surroundings much better than timid ones.

Select facilities that segregate cats from dogs and never put unfamiliar cats together in one kennel. More and more, "luxury" suites are becoming available that truly pamper your pet while you're on the road.

Relocating with Kitty

Over the past five years, my cats have moved with me five times. We've lived in houses and apartments in Florida, Pennsylvania, and California. From sprawling waterfront properties to tiny, temporary apartments with windows facing only one direction.

Yet, each time, within a day or so, my cats are playing and purring in the new place, proclaiming it as their new turf.

There are two means of refuge from the miseries of life: music and cats.
— Albert Schweitzer

What's my secret? I chat up the upcoming relocation with my cats before, during, and after each move. Strange as it may sound, I talk to my cats a few weeks before the move and let them know what will happen. Yes, strange men will be coming in and out of the house carting away furniture and belongings. I try to convey a sense of excitement and adventure. And, taking a lesson from the many realtors I've hired, I speak of the perks of the new place. Yes, there will be plenty of windowsills for you to perch on and bird watch. Yes, this one does have a set of stairs for your nightly workouts, and yes, it does come with an enclosed, screened porch.

Conveying a spirit of fun and adventure in your tone seems to help ensure smooth moves. Kittens that experience nonstressful moves will grow into adult cats that will adopt the feline blasé attitude of "been there, done that" if you should need to relocate again.

Q My kitty seems jittery because of all the moving boxes. How can I calm her down?

A Speak to her in soothing tones and treat her to extra massages to help her relax.

Many holistic-minded veterinarians also recommend adding a few drops of Rescue Remedy (Bach Flower Remedies, Ltd., Oxon, Engl.) — readily available at pet supply stores and drugstores — to your kitten's water, starting about two weeks before the big move. This homeopathic medicine contains a collection of flower essences that naturally help to diminish stress and instill a feeling of calm. (It's inexpensive and available at most health food stores and even most drugstores). At the same time, consider adding a few drops to your glass of water each day because kittens are savvy about reading your emotions. When you feel calm, it helps your kitten feel calm. Try it!

Q We're not moving far away. Is there anything I can do in advance to help my kitten feel at home at the new location?

A Curb your kitten's feelings of anxiety by rubbing a slightly damp towel on her back and then rubbing it on the walls, floors, and furniture inside the new home. When your kitten steps inside the new place, she will instantly recognize her own scent. She may even experience a sense of déjà vu — and feel more at ease in the new surroundings.

Q What should I do with my kitten during moving day?

A On the day of the big move, tuck your kitten inside an empty room (large bathroom or spare bedroom, depending on the locale). Post a large sign on colored paper that reads: KITTEN INSIDE. DO NOT OPEN! This will prevent overeager movers from barging inside and sending your frightened kitty fleeing to who knows where.

Stock this safety room with these items:

- A portable radio set on a pop-rock station to help drown out the sounds of moving
- A few favorite cat toys — such as catnip mice and a paper wad
- A scratching post
- Food and water bowls with a few treats
- A couple of your old, smelly T-shirts — great security blankets for kittens
- Litter box
- Pet carrier tucked in the corner

Your kitten should always be the last to be packed before you leave and the first to be unloaded when you arrive at your destination. Before the first box is unloaded from the moving van, rub your kitten with a slightly damp towel, then rub the towel on the baseboards of an empty room. Then stock this safety room as you did at home. The goal here is to try to duplicate the room at the old place on moving day. By providing familiar sights, sounds, and smells in a new place, you help your kitten adjust more quickly. I wait a few hours after the final boxes are unpacked before letting my cats explore the other parts of the home, one room at a time. Start with rooms with doors that you can close, and let them explore at their own pace. This step-by-step introduction builds their confidence and level of contentment.

Q I'll be moving soon and my kitten stubbornly refuses to wear a collar and tag — I don't want to lose her. What can I do?

A A smart solution is to book an appointment with your vet to get a tattoo for your kitty's ear or to have a microchip surgically implanted under her skin.

Your cat should be at least six months old before getting a tattoo. This unique alpha-numeric code is etched inside your kitty's ear where hair won't obscure its presence. All tattoo codes are recorded at a toll-free registry. See page 195 for details.

About the size of a grain of rice, a microchip contains a series of numbers or an alpha-numeric combination and is implanted under your cat's skin, usually between the shoulder blades. Don't worry — it doesn't go so deep as to enter the muscle and it is made of a material that won't irritate even the most sensitive of cats.

Fortunately, more and more animal shelters are being equipped with special detection wands so that they can check for identification of rescued cats not wearing collars

Changing Zip Codes

Add this to your moving checklist: update your kitten's ID tags before you move. On the day you get the new telephone number for your new locale, buy your kitten a new ID tag. Add it to kitty's collar and remove the old tag only after you've reached your destination. If your kitten happens to scoot out in the midst of the move, the people who find her will have a way to reach you.

and tags. They can then check a national computer data registry that lists information on pets having microchips.

Traveling by Air

Although the most common mode of transportation for felines is the car, some may earn their wings by traveling by airplane. See page 190 for an overview of pet policies by airline.

Q We're moving across the country. I've never moved with a kitten before. What should I do?

A Preparation and a relaxed attitude can help you avoid travel turbulence. Keep the following tips in mind to make your trip safe:

- **One month in advance:** Call the travel agent or airline to reserve space and get seat assignments. Book a nonstop flight when possible. Verify the airline's pet-carrier and pet-travel policies. Make sure your kitten's carrier is well ventilated, large enough for her to turn around in, and airline approved. The carrier should say so right on the label. Help your kitten adjust to this home-away-from home by taking her on short car trips inside the carrier.
- **10 days in advance:** Visit a veterinarian for a health certificate, physical examination, and any necessary vaccinations to keep your kitten on schedule. Tuck the health certificate inside the carrier pocket or keep it with your plane tickets so you don't lose it. Keep a separate copy in your luggage. If you know your kitten doesn't travel well, speak to your veterinarian about tranquilizers or relaxing herbs that may help her. *Note:* Some airlines require certificates 30 days in advance of travel.
- **Five days in advance:** Place a collar, identification tag, leash, and harness inside the carrier. Label the carrier with your name, home address, and phone number, and a contact name, address, and phone number in the city to which you are traveling. Take a photo of your kitten (or select one

from the dozens you've already taken) and keep it with your airline tickets. That's a safety measure just in case your kitten gets loose inside a big airport like Chicago's O'Hare International and you need to find her.

- **Twelve hours in advance:** Feed your kitten one last time, then withhold food. Pack a carry-on bag with a small bottle of water, dry food, and absorbent baby diapers in case of kitty potty accidents.
- **Before leaving for the airport:** Offer your kitten a small amount of water. Buckle your kitten into her collar and stash the leash and harness inside the side pocket of the carrier. Place your kitty inside the main compartment of the carrier and secure the carrier in the backseat of the car with a seatbelt.
- **At the airport:** Check in at the main ticket counter. *Never* remove your kitty from her carrier, especially at the metal detector. Ask the security person to use a wand to scan the kitten inside the carrier and allow you to walk through the metal detector with your kitten in tow. *Never* let your kitten be put through the X-ray machine. If the security person insists on removing your kitten from the carrier, ask to speak with a supervisor and explain your fear that your kitten might bolt out of the carrier and get lost inside the airport.

Q What can I do to make my kitten more comfortable on board the plane? She'll travel under the seat in front of me.

A Remain calm and relaxed. To soothe your kitten, pet her through the opening in the carrier. Don't draw a lot of attention to her. Some passengers may be allergic to cat dander and hair. Resist the desire to pull her out of her carrier midway through the flight to place her on your lap. Most airlines prohibit this, and you could spend the rest of the flight chasing her up and down the aisle. Once the plane lands, let most of the other passengers exit before you do, so your kitten won't feel harried. Once you're in a safe, confined area, offer your kitten some water and a tiny amount of dry food.

Safe Air Travel for Animals Act

In April 2000, former President Clinton signed the Safe Air Travel for Animals Act into law as part of the Federal Aviation Administration bill. The legislation requires airlines to provide monthly reports to the U.S. Department of Transportation Secretary on all incidents of loss, injury, or death of animals. The reports will be available to the public so that consumers will be better informed on specific airline track records regarding animal safety. The legislation also mandates improved training in animal care and safe transport techniques for baggage handlers.

More information about the Safe Air Travel for Animals Act is available from the Humane Society of the United States (HSUS). See page 194 for contact information.

Q What if I decide to transport my kitten in the cargo area of the airplane?

A Weather and regulations are your biggest challenges. Federal Aviation Administration rules prohibit the transport of cats or dogs during extremely hot or cold weather.

- Select airlines that equip their airplanes with ventilation in class D (cargo) — the compartment for transporting live animals and baggage. Airlines are not required to provide ventilation by law, at least not yet.
- If you're traveling during mild temperatures, call ahead to the airline to learn about specific requirements for pets traveling as cargo.
- Try to travel midweek in the evening when planes are less crowded. Choose direct flights whenever possible to avoid accidental transfers or delays.

There are many intelligent species in the universe. They are all owned by cats.

— Anonymous

- *Always* travel on the same flight as your kitten.
- Don't put any food in the crate. The ride will be upsetting and food can cause indigestion problems.
- Do put ice cubes in the water tray of the crate, which will melt and help keep your kitty from getting dehydrated.
- *Never* fly short-faced breeds such as Himalayans or Persians in the cargo hold. They may experience breathing difficulties and heatstroke.

Kitty Caper: Goodbye Sports Car, Hello Station Wagon

I loved my silver Datsun 287-2X sports car. It was smart looking, fast, and fun to drive. And the best part: I owned it. But then I got the opportunity to switch careers and move from Pittsburgh to Oceanside, Maryland.

The moving van took most of my belongings leaving me with a nine-hour road trip shared by my four young felines: Peaches, Pokey, Spencer, and Cary (as in Cary Grant). Concerned for their riding comfort, I sold my sports car for a roomier but used Volkswagen station wagon. Each cat was treated to a large plastic carrier, big enough to house a small portable litter box, water, food, and bedding.

I thought I was being the ultimate cat mom, sacrificing my beloved sports car to cater to their every need during this long trek. But guess what? During the entire nine hours inside the car not one of them ate, drank, or needed a potty break. Go figure.

—Karen Commings
Harrisburg, Penn.

Health Concerns

Take a moment to consider a veterinary visit from an impressionable kitten's perspective. First, there is the sheer indignity of being forced to hole up inside a tiny portable carrier no larger than your litter box. Then, there is that very unpleasant ride inside the car. Finally, the veterinary clinic door opens — and you're flooded with hideous smells, sounds, and sights. Egads! There are dogs here — big ones. Snarly ones. And, some of the cats don't look too friendly, either.

After barely surviving the waiting room, you're unceremoniously taken inside an enclosed room and placed on a cold stainless steel table that puts a chill to your paw pads. Some stranger in a lab coat starts poking and prodding your body, touching your fur, looking down your throat, putting instruments inside your ears and, ouch! Where did that needle come from?

Get the idea? Not exactly the dream family outing if you're a kitten new to the vet scene. Once you've recognized all the dangers — real and imaginary — to which your kitten may be exposed, it will be easier for you to take the steps necessary to make veterinary visits less traumatic for your feline friend.

Choosing a Veterinarian

It's time to select a doctor for your kitten. Word-of-mouth recommendations are helpful, but you'll want to do some investigating on your own, too.

Q How can I choose a veterinarian who will meet my kitten's needs?

A Choose a veterinarian as carefully as you would a car dealer and more deliberately than you would an ice-cream flavor. In other words, shop, look, and listen. Chat with your pet-owning friends and select a handful of potential doctors for your kitten.

Spend a little time interviewing prospective veterinarians on the telephone. You'll want to know answers to these questions:

- What are your office hours?
- Are you open evenings or Saturdays?
- Are you affiliated with an emergency clinic for after-hour care?
- How many veterinarians are on staff?
- Does the staff include licensed veterinarian technicians?
- Do you have veterinarians who specialize in cat care?
- If you prefer one-stop shopping, does the clinic offer grooming, boarding, and nutritional counseling services?
- If you find all previous answers satisfactory, then ask the make-or-break question: Can I stop by for a brief visit to see the clinic and meet the veterinarians?

Reputable clinics welcome such visits. Scratch off your list any clinic that is "too busy" to give you a five-minute tour.

Q What's the best way to evaluate a veterinary clinic when I make my initial visit?

A Don't bring your kitten with you on this visit. When you walk in the door to the clinic, survey the scene and ask yourself these questions:

- Is the waiting room crowded and noisy?
- What's the mood inside — calm and happy or rushed and stressful?
- Is there a double set of doors in the entryway to reduce the risk of a frightened kitten dashing into the parking lot?
- Are there separate waiting areas for dogs and cats?
- Does the place look and smell clean?
- Does the office receptionist greet you or ignore you?
- Are there photos on the wall of satisfied customers and letters of appreciation?
- Does the clinic have a separate lab area? A surgical area? A pharmacy?
- If it has boarding kennels, are the cages clean and are the housed dogs and cats quiet and calm?
- When you meet the veterinarians, do they take the time to answer your questions or do they appear harried?

Q What if I don't like the veterinarian after my kitten's initial visit?

A When you make a decision, don't ever feel as if you are locked into a lifelong commitment. If the veterinarian spends only a few minutes with your kitten on your initial visit and doesn't even call him by his name, make it your last visit. Go elsewhere to find a veterinarian with a "bedside manner" that ensures your kitten will receive personalized care.

Make sure the doctor communicates easily and is willing to listen to your concerns and to answer your questions.

Your kitten can live fifteen years, twenty years, or longer. He deserves a "family doc" who is willing to spend the time to get to know him, understand his needs, and keeps up-to-date on the latest medical advances.

Q Veterinary bills can really add up. Is it worth getting pet insurance for my kitten?

A No one can predict the medical maladies that may or may not develop in your growing kitten. Medical bills for cats with chronic conditions such as diabetes and hyperthyroidism can take a bite out of the monthly family budget.

Pet health insurance has been available since the early 1980s, but only 1 percent of pet owners take out policies, according to a recent survey by the American Animal Hospital Association. When planning your annual kitten budget, expect to spend, on average, about $130 for routine medical care and about $120 on food. When you factor in the cost for toys, spay/neutering, dental cleanings, grooming, and miscellaneous costs, the annual bill can total about $700, according to the American Pet Product Manufacturers Association National Pet Owners survey.

In considering pet insurance, be sure to interview the potential providers thoroughly. Before you decide, know what your deductible will be per year, if the policy follows a fee schedule, and if the policy covers basic wellness care or only accidents and illnesses. Know if the policy has payment limits, and shop around for prices and services. And start early: the younger the kitten the better. Insurance premiums are at their lowest when your kitten is at his healthiest.

Some veterinary clinics also offer wellness plans for cats, providing discounts for a package of services. It's worth asking to find out if your veterinarian offers such a deal.

Checkups and Vaccinations

Essential to your kitten's health is making sure he receives regular veterinary checkups and necessary vaccinations.

Q My kitten will need to make several veterinary visits for checkups and vaccinations during his first year. What can I do to make these experiences a little less dreadful?

A Some things in medicine are inevitable. A veterinarian must poke and prod a bit, open the mouth, check the eyes and ears and yes, give injections. It's all part of an essential physical exam for a growing feline.

The remedy? Practice preventive medicine at home. Do your part to desensitize your kitten by getting him used to having his fur touched, opening his mouth, lightly pressing on his footpads, and looking in his ears and eyes.

Every few weeks, conduct an at-home two-minute exam. For the exam, place your kitten atop your washer or dryer, which approximates the slickness and height of most exam tables.

- **"Weigh" your kitten.** Look for changes in body weight by standing above the kitten and looking for a slight "waist" behind his ribs. Then place both hands on the animal's ribs. You should be able to feel them, but they should not stick out. Check for fat pouches in the groin area between the hind legs and under the belly. If you have a scale with ounces, you can also weigh and record your kitten's weight.
- **Check your kitten's coat.** It should feel uniformly smooth. Check for dandruff, scaly skin, cuts, fleas, or ticks.
- **Feel for unusual lumps or bumps.** Move the hands behind the front legs, under the shoulders, down the back, over the hips, and down the legs. Inspect claws and footpads for cuts or cracks.
- **Gently pull down the lower eyelid to check for pink color.** The whites of the eye should be glossy white with no redness. Look for normal pupil size and responsiveness of the pupil to light. The pupils should get smaller in size when exposed to light. Watch for colored discharge, which can be a sign of infection.
- **Examine the ears.** They should appear clean, pink, and free of debris and strong odors.
- **Examine the gums.** Lift the lips away from the gums and press a finger gently but firmly over an upper tooth. When you remove your finger, the pink color should quickly return to the gums. Inspect all the teeth to make sure they are not chipped or missing.

These regular two-minute exams will give you a baseline of what your kitten looks and feels like. If you find anything unusual, contact your veterinarian. It's better to catch and treat a problem early than to wait until it becomes more serious.

Q My kitten is scheduled for his first physical exam. What does this entail?

A You're getting your new kitten off on the right paw by having him examined by a trained professional. When you enter the exam room, encourage your kitten to explore the examination table. Make it seem like a big adventure, not a scary experience.

You set the mood. If you're nervous or edgy, your kitten will pick up on your body cues and become nervous and edgy. If you're calm and upbeat, your kitty will think, "Hey, this is a new cool place to explore. My owner is right here with me."

While you are waiting for the veterinarian, introduce your kitten to the exam room. You might bring a spare shoelace and let him play a game of mini-chase to build up his confidence inside this new space.

A thorough physical exam usually takes between five and ten minutes. During the exam, your kitten will be checked from ear to tail and everywhere in between. He will be weighed to the ounce and have his body temperature taken with a rectal thermometer. His eyes will be checked for any discharge, swelling, or redness — all signs of a possible infection. The ears will be examined for cleanliness and the possible presence of ear mites, which look like old coffee grounds. The nose will also be scrutinized for any mucus, a possible sign of an upper respiratory infection.

Your kitten will have to open wide so the vet can look inside his mouth and check the condition of her gums, tongue, and teeth. The gums should be pink to the touch and teeth should not be chipped.

The veterinarian will listen to your kitten's heart and lungs with a stethoscope to detect any possible irregularities, such as a heart murmur or lung congestion. Your kitten's abdominal area is palpated to detect bloating, fluid buildup, or possible

FELINE FACT

A cat's heart beats twice as fast as a human heart, at a rate of between 110 and 140 beats per minute. They also have a speedier metabolic rate.

hernia. His coat will be combed to detect signs of fleas, ticks, scaling or hair loss. The paws, claws, rectum, genitalia, and tail are also examined.

Q I plan to keep my kitten indoors. Does he still need vaccinations?

A Yes. Even indoor cats occasionally slip out a door, so both you and he should be prepared. Certain vaccines help protect your young feline friend from many diseases and so are essential to his well-being. These vaccines teach your kitten's immune system to build up an army of antibodies that fight infectious invaders. Kittens receive a series of vaccines during their early months. Check with your veterinariana for the schedule that meets your kitten's needs and follows state laws.

Q What vaccinations do most veterinarians recommend for kittens?

A Vaccines should be administered based on a kitten's age, how many cats are in the household, if you plan to board your kitten, if he is strictly an indoor kitten or will go outdoors, and if he came from a shelter.

The American Association of Feline Practitioners and the Academy of Feline Medicine published their recommended guidelines in 1998. They advise that all cats receive a three-way vaccine to protect against feline panleukopenia virus, feline rhinotracheitis virus, and feline calicivirus, *plus* a vaccine to protect against the rabies virus, all of which are highly contagious.

Q When and how often should kittens receive vaccinations?

A The two feline medical organizations recommend giving the three-way vaccine (known as the FRCP vaccine for feline distemper) according to the following schedule:

- **First time:** when the kitten is between six and eight weeks old
- **Second time:** every three to four weeks until the kitten is twelve to fourteen weeks old
- **Third time:** when the kitten is one year old
- **Booster:** every one to three years depending on your kitty's health risks

The first rabies vaccine can be given to kittens that are twelve to fourteen weeks of age or older. The second rabies vaccine is usually administered when the kitten reaches one year of age. Then, depending on local state laws, subsequent rabies vaccines are given every one to three years.

Q What about vaccinations for other feline diseases?

A Other vaccines may be administered depending on your kitten's level of risk. Confer with your veterinarian so that together you can make a decision that best meets your kitten's health needs.

Other diseases that may be vaccinated against include:

- **Feline leukemia virus (FeLV):** Feline leukemia virus is the number one killing disease among cats. The virus attacks and suppresses the cat's immune system, resulting in cancer and other chronic and debilitating diseases. It is transmitted through direct contact with an infected cat, such as by mutual grooming, fighting or playing, or by sharing a food or water dish or litter box with an infected cat.

  This vaccine is recommended for cats that may potentially come into contact with other cats.
- **Feline infectious peritonitis (FIP):** This virus is fatal to cats. It can cause a cat to lose weight, vomit, have diarrhea, and experience neurological disorders. The method of transmission is unclear.
- **Chlamydia:** This bacterial infection triggers upper respiratory disease in cats. Like FeLV, it is spread by cat-to-cat contact.

- **Ringworm fungus:** Highly infectious, this condition can cause skin rashes and lesions.
- ***Bordetella bronchiseptica:*** This bacterium can result in upper respiratory infections and pneumonia.

Q Okay, I know needles are necessary for vaccinations. What can I do to keep my kitten from being scared?

A The worst reaction you can make as an owner is to scream, "ouch!" if your kitten cries a bit after receiving an injection, or immediately rushing to his side and saying, "Oh, that's okay. My poor little kitty." If your kitten sings out during the first vaccination and you heap on the coddling sounds, he may cry out even louder with the second injection, just to get the same level of reassurance from you. Rather than calming kitty, your reaction draws the kitten's attention to the injection. Instead, display an air of support and calm. Your kitten will take his cues from you.

Spaying and Neutering

Every day, about 22,000 cats and dogs are euthanized at animal shelters in the United States because they are not adopted. Yearly, the total ranges between 5 and 8 million, or about the population of the state of Virginia. Spaying and neutering prevents needless litters of homeless kittens.

Q Why is it so important to have kittens spayed or neutered?

A Unless you are a professional breeder who responsibly oversees the breeding of your cat, you should spay or neuter your cat. Altering an animal before he reaches puberty is the best way to avoid unwanted litters. Picture this if you dare: a pair of mating cats plus their offspring can produce 127,000 cats in only six years!

FELINE FACT

The percentage of spayed and neutered cats and dogs in the United States has increased by 400 percent since the first Spay Day USA in 1995.

Fees, on average, range between $45 to $160 to spay a female cat or dog and between $40 to $125 to neuter a male cat or dog. Prices vary based on the age and health of the animal, the area of the country, and whether the surgery was performed at a public shelter or private veterinary clinic. To locate a spay/neuter in your program, contact Spay/USA (see page 195 for details).

Yes, indoor cats also benefit from being spayed (females) or neutered (males). Let me tell you why:

- Altered cats, on average, live longer, healthier lives than unaltered cats.
- Females spayed before their first heat (estrus) cycle are at a dramatically lower risk for developing uterine infections and ovarian and breast cancer.
- Neutered males develop substantially fewer prostrate problems (including cysts, abscesses, and cancer) and have a no chance of incurring testicular cancer.
- Altered cats usually behave better because they are not driven by hormonal urges to escape outside to find a mate.
- Neutered males are less prone to get into cat fights.
- Thanks to medical advances, kittens can be safely altered as early as six weeks, providing that they weigh at least two pounds. They recover quickly, usually within a day or two.

Q How young can a kitten be to be spayed or neutered?

A Thanks to advances in anesthesia and surgical procedures, the sterilization procedure can be performed on a healthy kitten as young as six weeks old that weighs at least two pounds.

The recovery for this youthful surgical patient is extraordinarily quick, especially when the surgery is performed before six months or before a female's first heat, says W. Marvin Mackie, a veterinarian who heads four Animal Birth Control clinics in southern California.

"In view of the massive numbers of pets destroyed each year, I consider all of us in the veterinary profession to be part of a great failure to relieve pain and suffering by not advocating spaying before the female's first heat," he says.

Q What is done during spaying surgery and why will it benefit my female kitten?

A Spaying is the removal of the uterus and the ovaries. Therefore, heat periods no longer occur. In many cases, in spite of your best efforts, a female cat may become pregnant; spaying prevents unplanned litters of kittens.

Spaying offers several advantages. The female's heat periods occur usually twice a year, resulting in about two to three weeks of obnoxious behavior each time. This can be quite annoying if your cat is kept indoors. Male cats are attracted from blocks away and, in fact, seem to come out of the woodwork. Your cat would have a heat period about every two to three weeks until she was bred.

Spaying a female cat before her first heat sharply reduces her risk for breast cancer and uterine infections.

> **FELINE FACT**
>
> The gestation period for kittens is about sixty-three days. The birth, on average, takes about two hours depending on the size of the litter and the health of the mother cat.

Q If I spay my kitten, she'll never be a mother. Won't that make her sad? I'm sure I could find homes for the kittens.

A Let me dispel some myths. First, cats are not aware of the miracle of birth. They function instinctively, and probably would be much happier if they didn't have whining kittens to take care of. Second, you can't always count on friends to adopt a litter of kittens. Many may be filled with good intentions, but what if they live in an apartment where pets aren't allowed? Kittens also require lots of care, more than some people realize. Are your friends ready to make a lifelong commitment?

Q What are the benefits of neutering my male kitten?

A Neutering offers several advantages. Intact male cats go through a significant personality change as they mature. They become very possessive of their territory and mark it with their urine to ward off other cats.

A tomcat's urine develops a very strong odor that will be almost impossible to remove from your house. They also try constantly to defend their territory, which means one cat fight after another — another good reason to keep your cat indoors. Fighting results in severe infections and abscesses and often makes your neighbors angry.

You can often reduce hormonal spraying of urine by up to 90 percent by neutering your cat. Neutered males have minimal risk for hormone-related cancers as well.

Dental Care

Yes, when it comes to caring for your kitty, you need to be down in the mouth — his mouth. Don't overlook dental care as a part of your kitten's health care regimen.

Q What can I do to start my kitten on the path to good dental care?

A I'm sure you've noticed that high-spirited kittens have short attention spans. So, when establishing good dental hygiene habits, limit the first few training sessions to less than a minute apiece and a few teeth at a time. And don't wait until your kitten has a mouthful of adult teeth to start.

Fortunately, most kittens can be trained early to tolerate tooth brushing. After the kitten adjusts to the procedure, gradually increase the number of teeth you brush each day. Eventually your kitten may let you open his mouth wide enough to brush the inside tooth surfaces without gagging. Look Ma, no cavities!

Q What's the best technique for brushing my kitten's teeth?

A Until a kitten gets used to you opening and poking inside his mouth, try using your finger, a gauze sponge, or a specialized thimble-looking cap that slips over your forefinger to clean his teeth.

Use veterinarian-approved toothpaste formulated for cats. Gently pull back your kitten's lips with one hand, and brush in a circular pattern in strokes horizontal to the gum margin. Once you've built up trust, introduce the kitty toothbrush. As an enticement, soak the toothbrush in watery liquid from a can of tuna. (Kittens prefer a fishy flavor to mint toothpastes.) Brush at least twice weekly.

> **FELINE FACT**
>
> A kitten begins cutting teeth at two to four weeks of age and loses these temporary teeth to the permanent ones at close to four months of age.

Q My kitten refuses to let me use a toothbrush on his teeth. Any suggestions?

A If your kitten won't let you brush his teeth, consider the dental plan nine out of ten kittens prefer: the "tuna" toothbrush. Every few days, give your kitten a small cube of fresh raw tuna, not from the can but from the deli. Buy a slab of tuna about the size of a deck of cards. Tuck it in a resealable freezer bag and keep it in the freezer. Cut off a small cube about the size of your thumbnail, let it defrost five or ten minutes and then serve this sushi to your kitty.

The fish offers a double bonus: it's nutritious and it doubles as a toothbrush. As your kitten chews on the tuna, the fish massages his gums and removes surface tartar build-up.

Q My kitty is teething and I'm constantly shooing him away from chewing on the electrical cords. How can I stop him?

A Some cats like chewing electrical cords because the covering is chewy. This is not only a bad habit, but it is also

FELINE FACT

By the time your kitty reaches adulthood, he will have thirty-two teeth.

potentially dangerous. Give your kitten an alternative by growing ryegrass in a couple of different containers so your cat has grass to chew on. In areas where there are lots of cords such as around appliances or under desks, try using cord covers to keep your cats safe or wrap the cords with aluminum foil — definitely a texture cats detest. Also, bitter apple spray works well. Simply spray bitter apple on cords and your kitten will stay far away.

Grooming

Felines, by nature, are fastidious groomers. They love to preen and primp themselves. Sometimes, though, they need some human assistance.

Q My kitten seems to be a fastidious groomer. Do I still need to comb and brush him?

A Absolutely. Cats with longhaired and shorthaired coats reap plenty of benefits from regular combing and brushing sessions. Grooming decreases hairballs, controls fleas, keeps the cat cleaner, minimizes matting, allows the cat more freedom of movement, removes dead skin/fur, and degreases the coat. And if you groom your kitty regularly, you'll discover less flyaway cat hair on your furniture and your clothing.

Q How should I introduce my kitten to grooming?

A You've already mastered the first step if you treat your kitten to regular massages (see page 169). He is used to your therapeutic touch and he trusts you. For the first few sessions, run a baby brush over your kitten's coat to help him get used to the feel of the bristles. Then gradually switch to cat-specific brushes and combs.

Q What's the best way to groom my longhaired kitten?

A On the longhaired breeds, coat length can vary from two to six inches, and all require brushing every other day to keep mats under control. Use a steel wide-tooth comb and a pin-brush (a handled brush with long, rigid metal pins coming out of a rubber pad. Work in sections, parting the hair and brushing from the skin out to remove tangles and dead hair. Then pull the brush through to the hair ends to distribute the coat's natural oils.

Q During the winter, the inside of our home becomes dry. When I try to groom my kitten, we both get shocked by static electricity. How can I stop this?

A When it comes to your relationship with your kitten, the last thing you need is for sparks to fly. Dry heat inside your home during the cold months produces a buildup of static electricity that is released when you two come into physical contact.

Using a humidifier to add moisture to your home's inside air is beneficial for both of you. You can also stop "zapping" your kitty by wiping her coat with an antistatic dryer sheet before you begin grooming.

Pinecone Brush

Misplaced your kitty's grooming brush? In a pinch, you can use a pinecone to groom your cat. It removes the dead hair, aids in flea and tick removal and, best of all, it's free! Choose a pinecone that's dry and relatively sap free. Sap may stick to your kitten's fur and then be ingested as he grooms himself after the pineconing session, causing stomach upset.

Q What's the best way to groom my shorthaired kitty?

A Shorthaired kittens can be super shedders when not brushed and combed at least once a week. Grooming weekly also helps prevent hairballs caused by overzealous self-grooming. Use a slicker brush — a handled brush that has fine wire pins bent at a slight angle — to groom the undercoat. Always brush gently in the direction of the coat to remove any snarls. Brush from the body to the ends of the hair, and then brush slightly against the grain of the coat, holding the unbrushed hair back as you brush. This helps to remove any loose hair. Finish by brushing lightly in the direction of the hair growth.

Q What's the best way to detangle snarls or matted fur?

A Sprinkle a little cornstarch into these trouble spots. Cornstarch works naturally to help loosen tangles. Then try to work your fingers through the snarl to separate the hairs. Follow with a wide-toothed comb. For any stubborn tangles or matted fur, use blunt-tipped scissors to clip away problem areas.

Q If I have brushed my kitten, why do I need to comb him?

A Running a fine-toothed flea comb through your kitten's coat will help flush out any possible fleas or ticks that may not be picked up by brush strokes. If you find fleas, put them in a bowl of soapy water. Fleas can't swim and the soapy surface will keep them from escaping. The flea comb also helps you spot any scabs, scales, and dead hair. If you find a tick on your kitten, don't attempt to pluck it out. Instead, smear a dab of petroleum jelly on the tick and leave it alone; this will smother the tick. Within a day or two the tick will die and fall off.

Q I just brushed and combed my kitten and now he's standing in the sunlight grooming himself. Why?

A Hey, your kitty appreciates your efforts, but he's giving himself a quick tongue bath just to show you who is really in charge. Don't take it personally. It's a cat thing. Always finish every grooming session with a tasty food treat so your kitten learns to associate this primping time as a pleasing time.

Common Problems

Got a sick kitty? Let's review some common feline maladies and sensible solutions and treatment plans for each. When in doubt, always consult your veterinarian.

Gastrointestinal Problems

Humans don't have a monopoly on upset stomachs, gas, diarrhea, and other digestive disorders. Your kitty can also suffer from such unpleasantness.

Q What might cause my kitten to become constipated?

A If you notice that your kitten hasn't made a deposit in the litter box for a day, he's probably constipated and needs to be examined by a veterinarian. There are many causes. He may be sick and not exercising. Or he may have gotten into something he shouldn't have and eaten hair, bones, wool, or clothing. There may be an abscess around his anal sac or tumors that are causing painful defecation. He may be deficient in potassium, or dehydrated, or experiencing a side effect due to a medication.

EMERGENCY!

If your kitten displays a refusal to eat, acute diarrhea, repeated vomiting, swollen tongue, tender or painful abdomen, or convulsions, bring him directly to your veterinarian for immediate care. If you suspect your kitten may have nibbled on a houseplant before he got sick, tell the veterinarian what types of plants you own. It may save your kitten's life.

Q What can I give my kitten for occasional bouts of constipation?

A It's a cat fact: most normally have at least one bowel movement per day. If you scoop out the litter box and don't notice any deposits, try feeding your kitty some canned food or sprinkling a teaspoon of oat bran into his food bowl. Both should loosen his stools. Make sure your kitten has plenty of water to drink and, if necessary, use an eyedropper to feed him water. If two days pass and your kitty still hasn't had a bowel movement, have him checked by your veterinarian.

Q This is so embarrassing, but my kitten is a real gas, literally. Are there any herbs that can help with flatulence?

A Try to pinpoint what's causing your kitten's gassy episodes. The main culprit is often a diet that disagrees with your kitten's digestive system. Try gradually switching to another brand (see page 64), and see if the condition improves.

In the meantime, fennel, cayenne, chamomile, and ginger are top herbs that should help you "clear the air." Check with your veterinarian for the right dose and herb form for your kitten.

Q My kitten has diarrhea. What can I do to help him?

A If your kitten has a case of the runs — runny stools, that is — don't feed him for a day. If that doesn't relieve his condition, give him a bit of kaolin/pectin (Kaopectate). The recommended dose is one teaspoon per ten pounds of weight two or three times a day. Have your kitten examined by a veterinarian if the diarrhea doesn't disappear within twenty-four hours, and make sure your kitten is drinking plenty of water to reduce the risk of dehydration.

Sometimes the culprit behind running stools can be the table scraps you feed your kitty. Try adding fiber to your kitty's diet to firm the stool. Keep plenty of water for your kitten so he doesn't become dehydrated. To entice him to drink, make a broth of chicken soup and serve it at room temperature. Among herbs, slipper elm powder is the most effective for occasional diarrhea. If your kitty's diarrhea persists beyond two days, or you notice blood in his stool, bring him to your veterinarian immediately.

Q What's a hairball and how can I keep my kitten from having one?

A The medical term for hairball is *trichobezoar.* To your kitten, it means that during self-grooming, he ingests a lot of his own hair, which eventually finds its way into his intestines. Untreated, this hair accumulation can cause a serious intestinal blockage.

The simplest solution is to place a dab of petroleum jelly on your kitten's nose. He will automatically lick it off, ingest it, and the petroleum jelly will coat the stomach walls and help usher out hairball buildup. You can also put one teaspoon of fish oil or corn oil into your kitten's food once a week as a natural lubricant. Other hairball remedies include a spoonful of canned pumpkin or one teaspoon of bran daily.

Q Why is my kitten vomiting?

A Vomiting can result from any number of causes. Hairballs top the list.

If your kitten vomits within thirty minutes after eating, he may have a problem with the valve between the stomach and small intestine. This valve, called the pylorus, may spasm or be unable to open far enough to allow passage of food into the small intestine. Another possibility is an esophageal disorder that may prevent food from reaching the stomach in the first place.

To identify the cause, a veterinarian will take X rays of your kitten's thorax and stomach and then administer a barium dye to elucidate the esophagus, stomach, and upper small intestine. Work closely with your veterinarian to find a solution that will help your kitten.

Q Why is my kitten chewing my favorite wool sweater? There's a big hole in it.

A A kitten fond of eating wool or other fabric may be suffering from a condition known as *pica syndrome,* which is characterized by a compulsion to ingest nonfood objects. It's hard to say what causes pica, but experts theorize that cats are prone to this condition because they lack fiber in their diets, were weaned prematurely, or have separation anxiety or compulsive behavior.

Prevent this problem by kitty-proofing your home and taking away any possible temptation — specifically, your favorite sweaters. Store clothing in closed closets or laundry hampers, not on the floor. Drape and tuck a cotton bed sheet over the couch to prevent your kitty from munching on the fabric.

If you catch your kitten eating nonfood items, distract his attention. Schedule an appointment to have him examined by your veterinarian.

Infestations and Parasites

Itch, itch, scratch, scratch — your kitty is being bugged by bugs! Protect him with these tactics.

Q Why is my kitten constantly scratching himself?

A Frequent scratching may indicate that your kitten has fleas, allergies, or fungal infection, and the itch may be driving your kitty crazy. In his determined attempt to stop the itch, he may scratch his skin and yank out large patches of fur. Red patches and scabs may develop.

The most common cause is flea infestation and secondary flea-bite-allergy dermatitis. Infectious and allergic causes are also possible. Fungal infections, such as ringworm and certain types of mange, can cause itchy skin and may be contagious to humans and other pets.

To determine the cause of dermatitis, fungal testing and skin scrapings may be needed. If your kitten has more than an occasional itch, have him examined by your veterinarian. Unfortunately, not all itchy skin problems can be cured, but you can manage the problem to keep your cat content and comfortable. Some treatments for skin diseases include flea control, medicated shampoos, and special diets.

Q What can be done about fleas?

A Fleas are feisty and resourceful. They've been on the planet since the days of dinosaurs, so they also know a thing or two about surviving and adapting to a changing environment.

Keep in mind that fleas do not stay on your kitten all the time. They will jump off and seek another blood source or snooze inside the fibers of your carpet. Unchecked, fleas can quickly multiply. In fact, in one month, two dozen adult fleas can produce more than 250,000 offspring!

Check with your veterinarian about the best flea treatment plan for your kitten. A number of new once-a-month medications that are safe and effective on kittens as young as six weeks old are available in pill or topical droplet forms.

Q What are some helpful strategies to fight fleas in and around the house?

A Veterinarians recommend these healthy housekeeping habits to keep fleas at bay:

- **Vacuum your home weekly.** Don't forget the crevices of couches and chairs and cracks in basement cement floors. A vacuum cleaner equipped with a "beater bar" that is powerful enough to suck up adult fleas, larvae, and eggs is your best bet. Seal the bag and toss it outside in a garbage bin and cover it with a lid as soon as you've finished vacuuming.
- **Run a flea comb through your animal at least twice a week.** With each stroke, dip the comb in a bowl of hot soapy water or diluted rubbing alcohol. Fleas can't swim.

■ **Toss your pet's bedding in the washer on hot water weekly.** Include your pet's favorite rugs and towels to kill any fleas and eggs.

■ **Patrol your yard weekly.** Clear out sun-blocking vegetation near the home, rake up wet leaves and wet grass clippings, and do not store garbage under porches. All conditions attract fleas.

Natural Flea Remedies

Allow me to share with you a few flea-be-gone success stories. No longer should you or your cat be "bugged" by these pests.

Pedro and Michelle Rivera and their three spirited cats and four playful dogs reside in a rural setting on the outskirts of Madison, Wisconsin, far from city lights, fast-food drive-through restaurants, and bumper-to-bumper rush hour traffic. Their critters bound in and out the front and back doors frequently during the day, all year long.

"None of our pets has fleas. Not a single one," declares Michelle Rivera, a licensed massage therapist and herbalist. "They've been flea-free for years."

What's their secret? The Riveras attack fleas safely and naturally. Their natural lineup to make fleas flee includes homemade meals, vitamin and mineral supplements, specific herbs, and itty-bitty yard residents known as nematodes.

"We've found that the number one method to control fleas is proper nutrition," says Pedro Rivera, a holistic veterinarian. "Fleas are parasites that tend to live on unhealthy animals with low immune systems. "If an animal is healthy, it has a strong immune system and it won't harbor parasites."

■ **Scrub your pet's favorite chew toys, collar, and litter box pan in hot soapy water weekly.** This tactic prevents flea eggs from hatching. While washing the collar, be sure to put a back-up collar with ID tag on your kitty as a safety measure, just in case she accidentally gets out of your home and gets lost.

Mary Wulff-Tilford and husband, Gregory Tilford, both herbalists, rely on Mother Nature to prevent a flea infestation at their three-pet home in mountainous Conner, Montana.

They regularly sprinkle borate powder into the crevices of their couches and chairs. Two to three times a year, they stock their yard with shipments of flea larvae–eating nematodes that are available by mail-order or directly from some pet and garden supply stores. They add brewer's yeast and fresh garlic to their pets' food bowls as well as essential fatty acids such as omega-3 and omega-6. They brew tea made from freshly picked and chopped chrysanthemums and serve it as a liquid treat or flea rinse. They spray their window screens with a bottle of distilled water containing several drops of bitter orange essential oil; fleas hate citrus scents.

"Treating the entire environment, inside and outside, is critical," adds Mary. "People think that fleas spend most their time on animals, but they don't. They leap, feed, and leave and spend most of their time in the carpets, couches, and floors."

THE FORMIDABLE FLEA

Face it. The flea is a formidable foe that has been hopping about for centuries. Only by understanding what makes a flea tick, can you hope to win the battle. Lowell Ackerman, a veterinarian and board-certified veterinary dermatologist in Boston, shares these flea facts:

- Fleas can jump the equivalent of 150 times their size. To give you some perspective, that's like a person leaping 1,000 feet in the air!
- A flea rarely makes it past its first birthday, but don't drop your guard. Before it heads to flea heaven, a female flea can produce 600 offspring per month for a whopping total of 7,200 fleas during its brief life.
- Fleas prefer four-legged targets for their blood meals, but if cats or other animals aren't around, they'll choose you.
- Fleas love warm, moist climates. If given a choice, a flea will choose a climate with temperatures ranging between 65 to 80°F and 75 to 80 percent humidity.

Q What are ear mites?

A Ear mites are tiny insectlike parasites that live in the ear canal of cats and dogs. The most common sign of ear mite infection is scratching of the ears. Sometimes the ears will appear dirty because of a black material that looks like coffee grounds in the ear canal. The ears may be red or swollen or your kitten make shake his head excessively or paw at his ears.

Although they may leave the ear canals for short periods of time, ear mites spend most of their trouble-causing lives harbored in the ear canal. Ear mites are common in litters of kittens if their mother has ear mites.

Treatment usually includes a thorough cleaning of the ears by a veterinarian. Then medication, either topical or injectable, will be used to kill the mites.

Q Do all kittens have worms?

A Intestinal parasites are common in kittens. Kittens can become infected with parasites almost as soon as they are born. It may surprise you to learn that the single most important source of roundworm infection in kittens is the mother's milk.

Tapeworms are one of the most common intestinal parasite in cats. Kittens become infected with them when they swallow infected fleas. The eggs of the tapeworm live inside the flea. When the cat chews or licks his skin as a flea bites, the flea may be swallowed. The flea is digested within the cat's intestine, where the tapeworm hatches and anchors itself to the intestinal lining.

Cats infected with tapeworms will pass small segments of the worms in their stool. The segments are white in color and look like grains of rice. You may actually be able to see them crawling on the surface of the stool or stuck to the hair under the tail. Yuck!

If left untreated, these parasites can stunt a kitten's growth and cause vomiting and diarrhea. The microscopic examination of a stool sample usually helps to confirm the presence of intestinal parasites. Deworming medications can be given to a kitten as young as three weeks and repeated every three or four weeks. Deworming medication only kills the adult worms, which is why the repeat doses are necessary. Periodic deworming is highly recommended for cats that go outdoors.

Psychological Stress

We may fret about our hectic lives, but we aren't the only ones who can suffer the consequences of stress overload. Undue stress can harm your kitten's health, too.

Q My kitten looks like he has acne. Is that possible?

A Human teenagers aren't the only ones at risk for pimples. Feline acne usually appears as a localized infection of the chin area. Blame stress as the major cause of your cat's breakouts. Allergies can also trigger acne. Consult your veterinarian for the best treatment plan.

Q My kitty is a real scaredy cat. What can I do to calm his nerves?

A In extreme cases of behavior problems, sedatives and antianxiety medications may be necessary to help your kitten get back on an even keel emotionally. Check with your veterinarian or cat behaviorist to find a treatment program that addresses your kitten's needs. Medications should be used as a temporary measure until inappropriate behaviors can be modified.

Respiratory Problems and Allergies

Like humans, kittens are susceptible to allergies and other respiratory-related problems.

Q Can my kitten catch a cold?

A Ah-choo! Yes, he can. Upper respiratory infections, or kitty colds, are prevalent among cats. The most common causes are viruses. Acute viral infections may produce fever, congested nasal passages, thick nasal discharge, lethargy, and loss of appetite. Unfortunately — as for the common cold in humans — drugs that effectively kill these viruses do not exist, and veterinarians can only treat the symptoms. Occasionally, the primary viral infection may be complicated by a secondary bacterial infection, for which antibiotics may be given. Antibiotics relieve only bacterial infections.

Be aware that upper respiratory infections are extremely contagious. So, if you have a multicat household, the infection may pass from one cat to the next. An upper respiratory infection

> **CAUTION: PAWS DOWN TO ASPIRIN**
>
> Never give your kitten aspirin; his physiology just can't tolerate it. The same goes for acetaminophen. A single dose of either can kill a cat.

often takes seven to twenty-one days to run its course. If your kitten exhibits any cold symptoms, take him to your veterinarian to be examined.

The good news is that you can help your kitten feel better in small ways. Congested airways and sinuses can be relieved with vaporizers or humidifiers. Cleaning the discharges that build up around the cat's nostrils will help improve his breathing and sense of smell.

Many cats with upper respiratory infections have conjunctivitis. If the discharge around the eyes is watery and clear, clean it with a moist tissue. If the eyes are red, inflamed, and squinting, prescription medication is probably needed. Bring your kitten to the veterinarian.

If man could be crossed with a cat, it would improve man, but deteriorate the cat.

— Mark Twain

The best prevention? Annual vaccinations and reducing your kitten's contact with cats that do not live in your household.

Q My kitty is sneezing a lot — what's wrong?

A Your kitten may have an upper respiratory infection or a foreign object — such as a blade of grass — lodged inside his nose that may trigger the sneezing episodes. Allergies can also activate a series of sneezes. Have your kitten checked by your veterinarian if you notice a thick, nonclear nasal discharge or if your kitten seems depressed or has lost his appetite.

Q Can my kitten develop food allergies?

A Cats, just like people, may develop sensitivities to certain types of food. Signs of an allergic reaction may manifest as skin problems, specifically severe itching. Constant scratching to relieve an itch can lead to significant hair loss.

If you suspect that your kitten is allergic to something, work with your veterinarian or veterinary dermatologist to identify and eliminate the troublesome food allergen. Don't be surprised to discover that the culprit is an ingredient in your kitty's commercial food.

Home Care

You can be a great "home" veterinarian to your kitten. There are times when you must treat a minor cut or give him a pill or liquid medication. Here are some ways to make these occasions less traumatic for your kitten.

Q Sometimes my two kitties get rambunctious during play and accidentally cut one another. What's the best way to clean a simple cut or wound?

A Reach for a bottle of hydrogen peroxide and some distilled water. Make a mix of equal amounts of both liquids and dab it on the wound using a sterile gauze pad. Don't use hydrogen peroxide at full strength because it can burn new skin cells and slow the healing process.

Use the peroxide-water solution on the wounds twice daily, followed by a dab of topical antibiotic ointment. This regimen should stop the spread of infection and speed the healing.

Q Is there anything I can do to prevent an abscess?

A Anytime your kitties tangle, immediately check their coats for bloody areas or wet fur. Treat these areas to prevent infections. If a wound isn't cleaned properly, an abscess can form. Abscesses are infected pockets of fluid that may occur three to seven days after a cat bite and require veterinary attention. They need to be lanced surgically, drained, flushed, and treated with antibiotics.

Q My kitty has to take medication in pill form, but he's unwilling. What's the best way to give pills or capsules to my kitty?

A Most cat owners would gladly clean a dozen litter boxes a day before attempting to give one pill to their cat. It seems that cats may be psychic, able to predict precisely when it's pill-popping time. They take their cue and dash under the bed or run to some other hiding spot. Even if you do manage to catch them, they become contortionists and wiggle free before you can give them the pill. Solution: learn to outfox your feline.

- **Choose a small room.** Never attempt to give your kitten a pill in a wide-open area such as the kitchen or living room. Your kitten will surely disappear. Instead, give your kitten medicine inside a small room such as a bathroom. Shut the door. Speak calmly but confidently. If you have a wiggly kitty, wrap him loosely in a big bath towel, leaving only his head exposed. Once he realizes there's no escape, he'll give in and cooperate.
- **Method 1.** Now, you're ready to give the pill. The quickest method is to insert the pill into a ball of moist cat food and serve it as a treat. Be sure to follow this up with a soft treat to make sure that your kitten swallowed the pill.
- **Method 2.** If your kitten spits the pill out, try this. Open your kitten's jaw wide and pop the pill on top of his tongue as far back as possible. Then hold the jaws closed, tilt his head up slightly, and massage his throat to induce swallowing. Try blowing a quick puff of air into his face. When he blinks, he swallows — it's an automatic reflex.
- **Finish with praise and a tasty treat.** Ending on a positive note might make it easier for both of you next time.

Q How do I give liquid medication to my kitten?

A Use a plastic dropper; *never* use glass, which might break. Tilt your kitten's head slightly to one side. Place the dropper in the side of his mouth where the cheek pooch is and deliver the liquid in small but steady amounts. This pace makes your cat swallow each time. You can also deliver some medicine into the tip of his ears, using a dropper and then gently rubbing the liquid into the tiny capillaries of the ear for fast absorption. Check with your veterinarian before attempting this technique.

Always finish these medicine-giving ordeals with lots of praise and a tasty treat. Then, open the door and let your kitten scoot or saunter out, depending on his mood. Wait a few seconds and then slowly walk out and go in the opposite direction. He will quickly calm down and realize that you are not chasing after him.

Kitty Caper: Just the Fax, Ma'am

I work at home as a freelance writer and after I adopted Seren (short for Serendipity), I discovered that I kept missing important phone calls and faxes. Then I learned why. Seren, a champagne mink Siamese that was discovered homeless and sleeping in my friend's flowerpot on the back porch, had figured out how to play my "secretary" in my home office.

As a kitten, she would hook her paw under the receiver of my ringing phone and take it off the hook when I wasn't in the office. Who knows how many calls I missed because of this. Then, Seren would get my attention (always on a writing deadline) by leaping to the top of my fax machine. She learned that pressing all those pretty buttons made fun noises — *beep-trillll-llll-beep-pe-beeeeep!* Drove me crazy!

So now I've learned not to leave her unattended in my office. I've made it difficult for her to paw at my phone by setting it in a file tray with another file over the top of it. I placed a box that fits down over the top of my fax machine to keep her little paws from playing with the buttons. And, I've learned to make sure that my daily schedule — even on hectic days — includes some quality one-on-one time with Seren.

— Amy Shojai
Sherman, Texas

CHAPTER NINE

Alternative Therapies for Kittens

Planning to take your car-hating kitty on a long car trip? Even on short trips, your feline feels frantic, meows wildly, scratches at the sides of her carrier, and wants out *now.* Rather than make her groggy with a veterinarian-prescribed tranquilizer, the solution could be as simple as dabbing a few drops of Bach Flower essence oils (readily available at pet supply stores and drugstores) into her ears to calm her down naturally. Welcome to the age of alternative treatments for everyday kitten problems. More and more pet owners are seeking natural, safe, healthy treatment options that are free of the side effects often associated with conventional medications.

And why not? For their own aches and ills, pet owners are turning to the ancient healing arts of botanical herbs, acupuncture, and homeopathy. These holistic therapies work at healing both the body and the mind, treating the underlining cause, not merely the symptoms.

Still, as a kitten owner, if you're unfamiliar with herbal tinctures or the power of therapeutic massage, this can be a time of conflicting information and confusing choices.

Twenty years ago, medicinal herbs were only sold in natural health stores or by mail-order catalog. Today, you can buy healing herbs in pills, teas, tinctures, and other forms in your supermarket, drugstore, and pet food store.

Fortunately, more veterinarians are recognizing the value of an integrated approach to medicine; traditional treatments are paired with alternative ones such as herbs, acupuncture, and homeopathy. Membership in the American Holistic Veterinary Medical Association continues to grow each year as more veterinarians become trained in alternative medicine therapies.

Getting Started

Ready to pursue this healing journey for you and your kitten? Let's begin!

Q Help! Can you give me a quick rundown of the different types of alternative therapies available for cats?

A If you're new to the notion of holistic veterinary medicine, here are some brief descriptions of the main types:

- **Herbal medicine:** Plants have been used medicinally for thousands of years. Active constituents inside a plant's flowers, petals, stems, and/or roots can be used to prevent or treat a variety of physical and mental conditions. Herbal medicines work with the body's immune system to fight disease and improve the emotional and mental well-being of animals.
- **Homeopathy:** This discipline was developed in the late 1700s by German physician and chemist Samuel Hahnemann. He believed in the "law of similars," which states that like cures like. Homeopathy is based on the principle that a remedy in a large dose causes problems, but that a small homeopathic dose will stimulate healing. Homeopathic medicines are derived from plants, minerals, viruses, bacteria, drugs, and animal substances.

■ **Acupuncture/acupressure:** Both are based on relieving pain and improving the function of organ systems in the body by maintaining a balance of the body's energy life force known as *chi.* Acupuncture uses special narrow needles, and acupressure uses hand pressure to unblock specific energy-flow channels on the body called *meridians.*

■ **Chiropractic medicine:** Those trained in this holistic form of medicine know how to manipulate an animal's spine, bones, joints, and muscle to address physical problems, such as chronic pain caused by arthritis.

Q How can I find qualified veterinarians who practice alternative therapies in my geographical area?

A The good news is that the number of qualified holistic veterinary practitioners are growing steadily, and there's a good bet some are within driving distance of your hometown. Professional organizations that can assist you in locating some top-notch veterinarians in your area include the Academy of Veterinary Homeopathy, the American Holistic Veterinary Medical Association, and the International Veterinary Acupuncture Society. See pages 192–93 for contact information.

Q What should I look for in a qualified holistic practitioner for my kitten?

A Before you make a decision, interview a few practitioners in your area. Here's a handy checklist to help you decide which one will best suit your kitten's needs:

■ What are your qualifications in holistic medicine? Was it a weekend course or a 150-hour continuing education course that required you to pass a practical test for certification?

■ How long have you been practicing holistic medicine? What got you interested in it? What is your success rate?

■ Do you belong to a holistic veterinary medicine professional group and are you an active member?

- Would you be willing to provide names and telephone numbers of your clients?
- Are you willing to discuss my pet's case with the veterinarian who provides her primary traditional care?
- Will you explain holistic concepts to me and discuss any possible side effects or safety issues related to the holistic medicine or technique you recommend for my kitten?

You'll be paying for good service and the care of your kitten. Invest some time up front when choosing a practitioner and your kitten will benefit by having a health professional who looks after her needs.

Medicinal Herbs

Growing in your garden is Mother Nature's green pharmacy. Selecting the right herb in the right form and dose can yield therapeutic wonders for your ailing kitty.

Q What are the main forms of medicinal herbs?

A Giving herbal medicines to your kitten can be a little worrisome at first. But reaching for the right herb in the right form for minor mishaps and conditions can save you money and avoid a trip to the veterinary clinic. Medicinal herbs are available in these forms:

- **Fresh:** Snipped straight from the plant. Depending on the herb, you use its flowers, leaves, stems, or roots. Chop them finely and sprinkle them on top of your kitten's food. Fresh herbs are more potent than dried herbs.
- **Dried:** Although lacking the color, smell, and potency of fresh herbs, dried herbs are handy to have on hand when fresh herbs aren't available. The dried form has a longer shelf life than the fresh. Dried herbs should always crumble to the touch.
- **Teas:** Herbs that contain water-soluble ingredients work best in tea forms. Your kitten won't be sweetly sipping a cup of tea, but you can pour the herbal liquid over her food or

squirt the mixture along her lower back teeth with a syringe, available at your veterinary clinic or veterinary supply house. *Note:* Make sure you let the tea cool completely before giving it to your kitten.

- **Tinctures:** Also known as extracts, these concentrated and potent liquid botanicals are taken in small amounts — dropperfuls or teaspoonfuls. These herbal drops may be put into food or water or just dropped directly onto your kitten's tongue.
- **Capsules:** Herbs are usually pulverized and stuffed inside casings, often made from gelatin or enteric coatings, which speed absorption of the medicinal ingredients.
- **Salves and ointments:** These herbal creams are used externally on your kitten. Terrific choices to treat minor cuts and burns on your kitten.
- **Poultices:** Warm, moist masses of powdered or macerated fresh herb, which are applied directly to your kitten's skin to relieve insect bites and stings, inflammation, and blood poisoning. They work by drawing out infection, toxins, and foreign bodies embedded in the skin.

Q I take herbs to help fight off colds and to help me remember better. I wonder if my kitten would benefit from these natural plant medicines. What guidelines can you offer?

A When considering giving herbs to your kitten, Pedro Rivera, a holistic veterinarian who operates the Healing Oasis Wellness Center in Sturtevant, Wisconsin, says to keep these key points in mind:

- **Always consult your kitten's veterinarian or a professional herbalist before buying medicinal herbs.** If your kitten is on a prescribed medication, the herb could counter the effectiveness of the medication and worsen the condition.
- **Pay attention to the food you feed your kitten.** Proper nutrition can help avoid many illnesses. One good tip: avoid any products that contain synthetic food preservatives, such

as cookies, and processed foods, such as beef jerky. They may upset your kitten's tender tummy.

- **Treat botanical herbs like prescription drugs.** Although "natural" and derived from plants, they can cause sickness — even death — when used improperly. For instance, willow bark and garlic work safely on dogs but can be fatal in large doses to cats because of the physiological differences between the species.
- **Follow label directions.** Don't think more is better when it comes to herbal or homeopathic dosages. And, never give a human dose to your kitten.
- **Introduce herbal medicines one at a time.** Evaluate the results before adding others. Some herbs work well together; others clash.
- **Work with your veterinarian to wean your pet off prescription medications that can be replaced with herbal or homeopathic medicines, if possible.** In the long run, you could save money and your pet will face fewer side effects from the medications.
- **Don't expect miracles or instant results.** Natural medicine is wonderful, but it is not a cure-all. And, it should never replace diagnostic tests or, when necessary, antibiotics, anesthesia, or surgery. It should always be a complement to conventional medicine.

Kitty's Favorite Herb?

Without a doubt, catnip rules as the feline herb favorite. It provides added zip and zest in our cats, which, in turn, provide us with lots of amusing antics to watch. But only serve catnip to kittens that are older than six months. And, limit this cat-pleasing herb to only once a week until your kitten reaches its first birthday.

Q When I shop at the health food store, I feel a bit intimidated by the rows of different herbal products. Any shopping advice?

A There was a time when "shopping" for herbs meant a simple stroll into the woods to grab a handful of whatever plant you needed to meet your kitten's needs. Today, most of us are lucky if we have a patch of grass in our backyard. We must find our herbal allies in drugstores and pet supply stores.

To make a smart buy, consider these shopping tips:

- Limit your purchases to one or two herbs. Working with your holistic-minded veterinarian or an herbalist, take the time to see how effective these herbs are on your kitten before expanding into other herbs.
- Buy only certified organically grown products. This guarantees that no pesticides were used in harvesting or preparing the herbs for medicinal use.
- Only buy products that list both the common name (such as ginger) and the scientific name of the herb *(Zingiber officinale).* You need both to make the proper identification.
- Choose products that list dosage instructions. Keep in mind that dosages are typically based on an average human adult weighing 150 pounds. Check with your veterinarian to determine the proper dosage that will meet your kitten's weight and needs.
- Read the label and know what specific plant parts were used. The most medicinally active portion in some herbs may be in the roots; in others, it may be the leaves or flowers.
- Select products that list all ingredients in descending order of quantity, from the most active ingredients all the way down to solvents and fillers.
- Choose companies that list toll-free hotlines and Web sites and pack their products in double safety seal to prevent tampering and to keep the product fresh.
- Check for an expiration date, usually located on the bottom of the product's container. You risk losing the potency of the herbal product if it is expired.

Ten Herbs for Cats

Here are ten of the most versatile and safest herbs, in alphabetical order, that provide relief from many common cat conditions:

Common Name (Botanical Name)	Uses
Aloe *(Aloe vera)*	Minor burns, cuts, and insect bites
Calendula *(Calendula officinalis)*	Minor wounds, liver booster
Catnip *(Nepeta cataria)*	Initial euphoria followed by calmness; digestive tonic
Chamomile *(Matricaria recutita)*	Pain relief, minor wounds, anxiety relief
Echinacea (*Echinacea* spp.)	Immune system booster, infection fighter
Ginger *(Zingiber officinale)*	Improves appetite, relieves muscle pain, relieves diarrhea
Parsley *(Petroselinum crispum)*	Breath freshener, mild laxative
Red clover *(Trifolium pratense)*	Skin conditions, coughs
Slippery elm *(Ulmus rubra)*	Separation anxiety, pain reliever
Valerian *(Valeriana officinalis)*	Nervous stomach, minor wounds

Therapeutic Massage

Any kitten will tell you that petting is passé — what your cat wants, needs, and deserves are regular massages. Regularly massaged cats become accustomed to being handled. They associate touch with positive experiences. That can take the stress out of combing and brushing, nail trimming, car trips, veterinarian visits, and cat breed shows for both the cat and the owner, says Michelle Rivera, who operates the only state-certified animal massage school in Wisconsin. Sounds good.

Q Why are regular massages good for my kitten?

A This ancient therapeutic treatment does more than pamper. It keeps cats healthy, fit, and feeling fine. And, there are other benefits! Regular massage strengthens the cat-owner bond. Rubbing your kitten the right way should be viewed as a special time to pay attention and express affection for her. My once-aloof Callie practically stampedes my way whenever I say, "Callie, want a massage?"

Alicia Boyce, a veterinarian from Radicliffe, Kentucky, agrees that massage improves the bond between you and your kitten. She devotes ten minutes each morning to massaging her one-eyed cat, Earl, and her three-legged cat, Peg. She says it's a stress reliever both for her cats and for her.

Massage has medical merits, too. By moving your hands through your cat's coat, you can detect any lumps, cuts, fleas, or ticks. The earlier the detection, the quicker the cure! Massage can help relieve chronic conditions such as arthritis. Although not a cure for arthritis, massage reduces joint stiffness and pain by delivering oxygenated-rich blood to those trouble spots, says C. Sue Furman, Ph.D., an associate professor of anatomy and neurobiology at Colorado State University in Fort Collins, and owner of a canine and equine massage center.

Emotionally, experts say, massage strengthens the people-pet bond, helps curb aggression and other unwanted behaviors, and improves a cat's sociability with people and animals.

"Cats are very sensitive and because of this, tend to have emotional blockages," says Pedro Rivera, a holistic veterinarian in Madison, Wisconsin. With his wife, Michelle, he operates the Healing Oasis Wellness Center, which includes animal massage training for professionals. "Massage helps open up these blockages."

The cat always leaves a mark on his friend.

— Aesop

Q Medically, what happens to a kitten's body during a massage?

A Understanding how blood flows in your kitten will ensure that you perform the massage on your cat in a safe and therapeutic way. Massage strokes have very positive effects on the body when you understand how the cat's circulatory and lymph systems work, says Dr. Furman.

Dr. Furman provides this mini-lesson: The arteries in the heart pump out fresh, oxygenated blood to the body tissues. This pumping action enables the blood to move through arteries into smaller vessels and eventually into tiny capillaries where red blood cells pass in a single file.

These tiny vessels pick up and get rid of carbon dioxide, lactic acid, and other waste products from the body. The result? Improved circulation.

Now you know why your kitty starts purring when you start the massage session!

A Word of Caution

"A person needs to know a cat's anatomy and recognize that a technique that can be used on a person, dog, or a bird may not be ideal for a cat," says Michelle Rivera, who operates the only state-certified animal massage school in Wisconsin. "In your attempt to help, you could do more harm than good." Learn feline massage techniques from someone certified in animal massage.

GOING THROUGH THE MOTIONS

Following are six motions recommended by licensed massage therapist Maryjean Ballner. Practice these on a pillow or stuffed animal *before* trying them on your kitten:

- **Go with the glide:** This classic massage stroke is simply a straight, flowing, continuous motion. It usually moves from the top of the head, down the back to the tail.
- **Create circles:** Move your fingertips in clockwise or counterclockwise motions that are the diameters of half-dollars.
- **Do the wave:** Make side-to-side rocking strokes with open palm and flat fingers. This mimics the motion of a windshield wiper.
- **Focus on flicking:** Pretend that you are lightly brushing imaginary crumbs off a table and you've got the idea behind this motion. You can flick with one, two, or three fingers.
- **Here's the real rub:** Move your hand slowly along your cat's body, exerting featherlight, light, and mild pressures.
- **Heed the knead:** This gentle caress uses the flicking motion of your palm and all five fingers. Ideal for working the spine area.

Q How hard should I press when massaging?

A Never press too deeply, and always stroke the muscles in the direction of the heart to improve healthy blood flow. Use an airy touch, light caress, or mild strokes. Detour around an area of recent surgery or an open wound. Gently massage above and below these areas to stimulate blood and nutrient flow to speed healing.

Q How fast should I massage and for how long?

A Match the speed of your movement to the mood of your kitten. Let her dictate the tempo. A kitten just waking up from an afternoon snooze may prefer gentle, relaxed strokes. A kitty just finishing a frisky playtime, may seek faster, firmer strokes. Limit the massage session to five to ten minutes or less, if your kitty begins to act restless.

Q When is the best time to massage my kitten?

A Felines aren't shy about receiving therapeutic body rubs, but you need to let your kitten pick the idea time. Don't try to force a massage just to meet your schedule. Cats can read your body cues and know when you feel stressed or harried.

Step into your kitty's paws for a second. You're waking up from your third afternoon nap of the day and starting to do a yogalike stretch just as your owner bounds your way wearing that silly grin.

Now, still imagining yourself as your cat, which of the following would you prefer?

- **Scenario 1:** [Loud, excited voice] "Hey, Murphy! Great to see you! Come over here and get some petting!" These shrieking words are followed by a quick series of flat open palm thumps pounding on top of your head. Your owner views it as a friendly greeting, a pat. You view it as the start of another tension headache. *Or,*
- **Scenario 2:** [Softly] "Glad to see you're waking up from your snooze, Murphy. Would you like a little kitty massage?" You as the cat, stand up, stretch, and agree to a five-minute muscle massage that runs down the spine from your neck to the base of your tail. You feel marvelous.

The second scenario is preferred by most kittens.

Q My kitten seems to be in the mood for a massage. How do I begin?

A Make sure your hands are clean and dry. Then approach your cat slowly (this is no time for a foot race) and speak to her in a calm, soothing voice.

Spend a few minutes *gently* and *slowly* stretching your cat's limbs — one at a time — to warm the muscles and increase range of motion. Use your fingertips and your open hand, never your fingernails. Read your kitten's body language.

- **Good cue:** She loves the massage if she gives you a sleepy half-eyed look, noses you (a cat kiss), or even falls asleep in your hands.
- **Bad cue:** She would rather be elsewhere if she begins to resist, wiggles, gives you a full-pupil glare, and cries, *m-e-o-w!*, the cat equivalent of "Stop!" End the session and try again when your cat is in a massage mood.

Q Where are the best places to give a massage?

A Depending on your kitten, popular choices include in your arms, on her favorite blanket at the foot of your bed, on a wide windowsill, on the sofa or chair, or even on the carpeted floor.

Kitty Kaper: Life-Saving Massage

I heard from friends about giving their cats massages and decided to try it on my six-month-old kitty, Annie. As I glided my fingertips down her torso, I felt a pea-sized lump. I immediately took her to the veterinarian. Tests confirmed a cancerous tumor on Annie's mammary glands. The lump was completely removed and Annie is now eight years old and very healthy. Massage saved my kitten's life!

I now provide free massages to sheltered cats at the Vero Beach Humane Society. Massages improve the cats' moods, which increases their chances of being adopted.

— Laurie Iodice
Vero Beach, Fla.

It's Showtime: Teaching Your Kitty Tricks

Yes, you can teach your kitty tricks! Living proof: Murphy, my one-year-old sable-colored cat. Though she meows instead of barks, she displays a puppy enthusiasm for walks on a leash. She also sits up for treats, comes when I whistle, and lies down when I snap my fingers.

Maybe it was because I forgot to tell Murphy that she is a cat. Maybe she knows and decided to play along. Regardless, she represents the growing litter of kittens willing and able to master tricks that fit their fancy.

When it comes to teaching kitties tricks, think fun and think timing. Your mood and when you attempt to introduce a new trick play major roles in your level of success. Will your cat become a Hollywood star because he can come, sit, stay, fetch, or sit up on your command? Maybe. Maybe not. But the rewards for you and your kitty are priceless.

Q What's the payoff for my kitty and for me? Why invest the time in training?

A If you spend time training your kitten and teaching him some tricks, you will stimulate his mind, improve his self-confidence, sharpen his socialization skills, and reduce the likelihood of behavioral problems. The level of trust between the two of you will increase, and he will seek you out and want to be with you. When he sees you approach, his tail will go up in a friendly "Hi!" salute. He will be less of a behavioral problem and more of a feline pal.

Q What basic guidelines do I need to keep in mind when I start training my kitten?

A Glad you asked. Here are some pointers that will make the training easier and more enjoyable for both you and your kitten:

- Train when your kitten is receptive to learning — ideally, before meals when kitty is hungry — not simply when it fits your schedule.
- Select a quiet room where you can be one-on-one with your kitten.
- Always say your kitten's name first to get his attention.
- Be positive, patient, and encouraging.
- Be consistent with verbal commands and hand signals.
- Start with the basic commands: "Come," "Sit," "Stay."
- Gradually introduce advanced commands such as "Go fetch your mouse," or "Go to the kitchen door if you want to go for a walk outside."
- Immediately provide food rewards and praise for each success, no matter how small.
- Use your kitten's natural curiosity to your advantage and consistently let him know that you're out to pamper and please him.
- Make sure that your kitten succeeds. If he isn't performing the desired behavior, chances are good that you are moving too quickly. Make sure that your cat understands each training step before progressing to the next one.
- Be flexible, and recognize that sometimes a kitten just isn't in the mood to perform a trick.
- Teach your kitten only one new trick or behavior at a

time. Kittens are not masters of multitasking. Keep it simple and short — work no more than ten or fifteen minutes per session.

Simple Cat Tricks

Dogs don't have a monopoly on basic obedience. Kittens can master some basic tricks, too. Let's start with these:

- Come
- Silent hello
- Talk back
- Sit
- Stay
- Sit up
- Fetch

Come

Among cat owners, there is a popular saying: "Dogs come when they're called; cats take a message and get back to you later." But it doesn't have to be that way. You just need to be cunning with your kitten. This trick comes in handy when you need your cat to be within your reach. Always reward his compliance with praise or food. *Never* use this command to punish him for missing the litter box or scratching the couch.

The Can Opener. Recognize the motivators you have at your disposal. There seems to be a magical bond between kittens and electric can openers. Use this connection to your advantage. Your kitten will quickly associate the whirring sound of the can opener with the tasty treat that follows. He'll come running every time.

Use this conditioned response to teach your kitten the "Come" command. The best time to teach your kitten to come when you call him is at mealtime. He is highly motivated to pay attention to you at this time because he wants to eat.

The Tap or Click. When your kitten is in a room other than the kitchen, tap his empty food bowl with a metal spoon or click a child's clicker toy or turn on the electric can opener.

Say, "Whiskers, come here." Then, tap the bowl or make the clicking noise a few more times until your kitten reaches you. Your kitten, sensing that food is on its way, should be racing toward you. When he does, praise him and serve him a filled food bowl.

Repeat this ritual before each meal and he will quickly associate these sounds with getting a full belly.

The Whistle. If your kitty ignores the "Come" command, try teaching him to scurry your way by whistling and enticing him with food. Call your kitten by name, and then let loose with a "come-here" whistle. Tap on his food bowl if necessary to get his attention. When he arrives, praise him and give him a tasty meal. In no time, your kitten will associate food with your beckoning whistle.

My cat Murphy is so well conditioned that she comes to my whistle even if she is in a deep afternoon snooze. The whistle command came in handy the time she slipped out of my patio door, climbed up a tree, and was walking on my neighbor's barrel-tile roof. A few whistles were all it took for my adventurous, food-motivated Murphy to shimmy down the tree and calmly walk back in my house door to receive a tasty treat.

Cats are smarter than dogs. You can't get eight cats to pull a sled through the snow.

— Jeff Valdez

Her favorite whistle tune? Strange but true: the theme song from *Jeopardy!* Works every time — even in front of houseguests, unfortunately.

The Silent Hello

With your contented kitten within arm's reach, kneel down and make a loose fist with one hand. Slowly position your fist so that it is slightly extended straight in front of you at kitten eye-level. Be patient and let him approach you with a rub or a head butt. This is a way for him to greet you. It is also a good trick to master because it makes cats feel more at ease at being petted or picked up.

Talk Back

Position one of your kitten's favorite food treats a few inches from his face. Let him get a good whiff of this must-have morsel. Then speak his name a few times. When he answers back, give him the treat. To reinforce this cat chat, keep a few pieces of food treats in your pocket. Greet your kitten with a friendly "Hi," and each time he responds, praise him and give him a treat. He will soon make the association between meowing replies and treat delivery.

This trick comes in handy if your kitten ever sneaks out of your house or slips off a leash. He will be more apt to respond to your calling his name because of the treat training sessions. He may be hiding in a bush but let out an "I'm here" meow because he recognizes your voice.

Sit

If you've successfully taught your dog to master the "Sit" command, you can teach your kitten, too. Kitten see; kitten might do. This is an important command for your kitten to learn. Once a cat can sit on command, he is more receptive to learning other tricks and behaviors. Attempt this command just before your kitten's mealtime.

1. Select a quiet place where he feels comfortable and safe. Gently place him on a table near the edge closest to you. Give him some friendly pats so that he feels at ease.
2. Say, "Felix, sit" as you move a food treat slightly above his eye level and directly over his head.
3. When he tips his head back to follow the treat with his eyes, he'll need to sit down to maintain his balance. When he sits, click or snap your fingers and say, "Sit, good sit." Immediately give the food treat.
4. Repeat these steps until your cat obeys the "Sit" command with only the hand gesture, not the treat.
5. If he doesn't sit on his own, gently press down on his hindquarters. Click or snap your fingers and say, "Sit, good sit." Be gentle and patient so you don't frustrate or frighten your cat.

Once your cat has got this behavior down, practice the command when your cat is walking on the floor. How did he do?

Then bring in your obedient dog. Have your dog sit on command and give her a treat. Then do the same for your kitten. He definitely won't want to be shown up by a dog!

Stay

No one likes to have to chase a fleet-footed feline around the house to get him to put on his collar or to feed him a pill. Kittens are too quick, too agile. Instead, teach him the "Stay" command. Here's how.

1. Begin in an enclosed area, such as a screened porch or den, any place that doesn't offer him hiding spots or escape routes.
2. When kitty starts to move away from you, say his name followed by "Stay." Extend your arm straight out and, with the palm facing down, move it down steadily toward the floor. Maintain your standing position and don't move. Don't chase after your kitten.
3. Only after he sits or lies down should you slowly approach him, kneel down, and reward him with "Good boy, Felix" or "Way to go, Felix."
4. Pick him up and give him a friendly hug, his reward. Then put him back down on the ground. Repeat a few times.

Once he heeds the "Stay" command inside the enclosed room, practice the command in an open area like the kitchen, and say goodbye to your cat-chasing days.

Sit Up

Think of the "Sit up" command as the feline version of the classic dog beg-for-food trick. Cats are certainly too dignified to beg, so they prefer to sit up instead.

1. When your kitten appears calm and happy, gently place him in a sitting position. You can hoist him up on a sturdy chair or let him stay on the floor.

2. Hold a treat an inch or so over your kitten's head and say, "Sit up." If he tries to swat at the treat with his paw or stands up on his hind legs and grabs your hand, don't give him the treat.
3. Repeat the command "Sit up" while holding the treat over his head. When he sits up and balances his weight over his hind feet, reward him with the treat and praise him.
4. Repeat the command-behavior-reward sequence several times to reinforce the behavior.

When it's treat time at my household, Murphy and Little Guy rise up in perfect harmony.

Fetch

Cats are born predators. They love to chase, stalk, and capture. Tap into these instinctive drives when you want to tutor your kitten in the fine art of fetching. Select a large, uncluttered room or a long hallway for this trick.

1. Take a piece of paper and wad it into a ball about the diameter of a nickel; this seems to be a tempting size for most kittens. Make the ball in front of your kitten — the sound, sight, and odor of a fresh paper ball can be almost irresistible for most kitties.
2. Show your kitten the prized paper wad and toss it over his head. As you do, say, "Fetch."
3. Praise your kitten as he chases after the paper wad, bats it around, and grabs it in his paws or mouth.
4. Wave your hand to motion your kitten to come to you as you say, "Come here." You may need to retrieve the paper

wad a few times until your kitten understands how to play this game.

5. Reward kitty with a treat only when he brings the paper wad within a foot or so of your feet.

Advanced Cat Tricks

Think you've got a really smart kitty? One that craves a challenge? Perhaps he can master any or all of these more difficult tricks:

- Shaking hands
- Paw touch
- Rolling over
- Jumping up
- Walking on a leash
- Using a kitty door

This is just a sampling of sophisticated tricks. You are limited only by your imagination — and your kitten's willingness to cooperate. Now, if you teach your kitten how to switch channels on the television remote or how to fetch your car keys, let me know *your* secret. I'd be more curious than a cat to learn how you accomplish either of these feats!

Shaking Hands

Cats, by nature, are very front-paw oriented. Soon your kitten will be primed to greet like a politician on a campaign trial.

1. With your kitten positioned in front of you, touch his front paw with a small treat and say, "Shake."
2. The moment he lifts his paw, gently take it in your hand and shake it. Heap on the verbal praise and give him a small food treat.
3. Repeat these steps in sequence for four or five times. Stop once he delivers a couple of paw shakes or decides he is bored. Always keep training sessions short so your kitten has fun and doesn't regard them as chores.

Paw Touch

To accomplish this trick, your kitten must have mastered his basic feline obedience training and aced the basic tricks, especially the "Sit" command. You'll need treats and a clicker for this one.

1. Place your kitten on a sturdy, stable table, about twelve to fifteen inches from the edge closest to you. Instruct him to "Sit."
2. Place a small toy, thick book, or another object that won't fall or tip over when touched at the edge of the table, between you and your kitten.
3. Hold a small food treat in front of your kitten at the edge of the object nearest you. (The object should be between kitty and the treat.)
4. Say, "Felix, paw" as you touch the object with your free hand.
5. When your kitten reaches for the treat with one of his front paws and touches or steps on the object, click and say, "Paw, good paw," give him the treat, and praise him.
6. If he doesn't touch the object, try moving the treat from side to side to entice him to swat at it. As soon as he touches the object, click and say, "Paw, good paw," give him the treat, and praise him.

Once your kitten has performed this several times successfully over the span of a week, repeat these steps but don't use the clicker cue. Continue with the food treats. The final challenge? Say the command, "Paw, good paw" while pointing to an object on the ground. If kitty does this on cue, success! It doesn't matter whether he touches the object with his left or right paw or alternates.

Rolling Over

The good news is that if your kitten can flop on his side, he's halfway to the roll over. Move over, Fido! Any food-motivated kitten can learn the art of rolling over. Plus, this is a fun trick to have your kitten do for visiting friends.

1. When your kitten is sitting on the floor, kneel down in front of him.
2. With a treat in your right hand, slowly pass it over his left shoulder to the point that he must turn his head to look at it. Say, "Roll over" as you keep moving it up and over.
3. When kitty tries to grab the treat with his paws, he'll go belly up and roll over on his side. Reward him with the treat and praise.
4. Repeat a few times. She'll be flopping over on command before you know it.

Jumping Up

If your kitten scurries under the bed beyond your reach, you're in good company. Many shy kittens and cats, including my Callie, dash under the bed whenever the doorbell rings or when they see their owners pull the cat carrier out of the garage. They associate the doorbell with a visit by a stranger and the carrier with a trip to one of their least favorite places — the veterinary clinic.

Trying to coax your kitty out from under the bed can be nearly impossible. In this situation, the "Jump up" command comes in handy.

1. Shut closet doors and drawers in your bedroom.
2. Place kitty in your bedroom and close the door behind you both to prevent to block out any distractions and to prevent kitty from escaping. Most likely, your kitten will go directly under your bed.
3. Look under the bed and make eye contact with your kitten.
4. Tap the top of your bed with an open palm, while calmly but firmly saying, "Felix, jump up."
5. If necessary, coax your kitten out from under the bed by using the long end of a broom handle, but do not poke or prod him. Just use a gentle sweeping motion beside your kitty. Be patient but persistent. It could take several minutes for him to jump up on the bed.
6. When he jumps up on the bed, immediately give him a food treat and lots of praise. Let him know that you are proud.

7. Leave him on the bed and walk out of the bedroom, leaving the door open. This lets him know that he is free to stay or leave on his own terms.

Walking on a Leash

Face it. The world of indoor kittens is limited to the size of the house or apartment in which they live and what they can see outside by perching on windowsills. Training your kitten to walk on a leash gives him a safe, and chaperoned, passage into the great world outdoors.

Felines at any age can be trained to walk with leashes if the proper steps are taken, says Bonnie Beaver, a veterinarian specializing in animal behavior at Texas A&M's College of Veterinary Medicine in College Station.

"It could take a few days to perhaps a week to train a cat, depending on the personality of the cat and how often the owners try the lessons," she says. And always give food rewards. "Certainly praise is fine, but for cats, food is the big reward," says Dr. Beaver. "They are strongly motivated by food." Once your kitten understands that the harness and leash are his friends, you're all set. Remember to use treats sparingly; if kitty's no longer hungry for treats, he may lose interest completely.

1. Buy a cat harness that fits the barrel of your cat's body. The harness shouldn't be too snug or too loose. And do keep fashion in mind. Choose a color that doesn't clash with your kitten's furry coat. No dignified calico wants to sport a neon-pink harness in public.
2. For a couple of days, keep the harness and the leash in the toy box housing your cat's prized toy mouse, the catnip ball, and the old shoelace. Let him sniff and paw it without any interference from you. Because the leash and harness are in with some of his favorite things, he will begin making positive associations with them.

3. Place the harness on your kitten for about a minute — without attaching the leash — then give him a treat. Then remove the harness.
4. Gradually build up the time he wears the harness and always give him a treat before you remove it. Playing with kitty while he's wearing the harness will improve his comfort level.
5. Next, introduce the leash. Attach the leash to the harness and give your kitten a treat. Praise him. Flatter him. As he struts around the living room with the leash trailing behind, this is your cue to mew a few *oohs* and *aaahhs,* and sprinkle in more tasty rewards.
6. Cradle your happy, harnessed, leash-wearing cat in your arms and walk outside. Speak calmly as you point out some of the scenic sites like the robin on a low branch. Then bring him back inside and give him another treat. Great job. That's enough for one day.
7. Next time, take your leashed and harnessed kitten outside and gently lower him to the ground. Kneel down beside him and speak encouragingly. Let him choose where he will go.

Scout the Terrain

Before you take your kitten out for a walk, make sure that there are no dogs roaming. The outdoor walk should be a time of adventure for your kitten, not a time of terror. Also, keep your kitten's claws trimmed, so in the event that you have to scoop him up out of harm's way, he won't dig deeply into your skin.

Thousands of years ago, cats were worshiped as gods. Cats have never forgotten this.

— Anonymous

Leash training offers a fun, safe outdoor experience for you and your kitten. He gets to explore new scenes and walk off some calories. But don't expect your kitten to trot or to walk long distances like dogs. Kittens tend to take a few steps, then stop to sniff a flower, eye a bug crawling on the sidewalk, or perk their ears into the wind to listen for the approach of any possible foes. "Generally, cats have an innate curiosity. They go where they want and usually they walk you," cautions Dr. Beaver.

Using a Kitty Door

A cat-sized door is good option if you have an enclosed patio or porch. It permits your kitten to come and go at will from the main living space into an enclosed space, giving him a sense of the great outdoors in a safe, controlled setting. Choose a door that's appropriate for your and your kitten.

- **Select a door suitable for your home.** Door models are available for installation in existing doors, in walls, and as separate panel extensions for sliding glass doors.
- **Size the door.** The door should be slightly taller and wider than what your kitten will be at adulthood. Check with your veterinarian if you're unsure.
- **Consider your climate.** A non-toxic, plastic flip door performs well in mild climates but not in extreme cold or heat.
- **Choose a locking system.** Lock types include magnetic, latching, and electronic, and are typically mounted on a panel that conceals the kitty door.

Training your kitten to use the kitty door may take a day or a week; it all depends on your cat. As for all the other tricks in this chapter, be patient with your kitten and yourself, and use lots of treats.

1. Introduce the door to your curious kitty by luring him closer with some tasty treats. Give him a few days to sniff it out and gain confidence.
2. Detach the flap or door.
3. Have a friend stay in the house with your kitten, and go outside the door. Call your kitten by name and say, "Come." Hand out treats and heap on the praise each time he ventures through the opening. After a few successful trips, reattach the flap or door.

4. Repeat the "Come" command, and reward with treats and praise if he makes it through the door. Limit each training session to no more than fifteen minutes.

Parting Advice from a Pro

You might say that Rose Ordile's job is the cat's meow. She is a professional animal trainer in charge of the newest "Morris the Cat" for Heinz's 9-Lives Cat division.

The big orange tabby took over for a retiring Morris in the spring of 2000, becoming the fourth cat to serve as the company's feline ambassador. Not bad for a cat of humble beginnings. Ordile spotted the new Morris as a homeless stray at an animal shelter in southern California.

One look and she knew this cat had the savvy to be a celebrity cat comfortable in airplanes and in front of cameras and audiences. And, he possesses just the right charm, style, and wit to carry on the thirty-two-year-old tradition as Morris. Ordile adopted him and began the training once he got comfortable in her Canyon Country, California, home.

"I helped the 9-Lives team search hundreds of shelters looking for just the right new Morris," said Ordile. "I saw him on Valentine's Day, on the second level of cages sitting close to the front of his cage. He was a little aloof but social. Once I put him on the ground, he rubbed against my legs and starting walking up to the cats in other cages as if to say, 'Nya, nya, nya, I'm getting a home.' We needed a confident cat who could handle a lot of distractions, and this one was perfect."

Ordile uses positive reinforcement and conditioned response techniques in her training with Morris and other celebrity animals. She started feeding Morris in a studio chair, keeping his bowl there at mealtime. The blue chair features the 9-Lives logo.

"I taught him to stay in this chair and he has learned that only positive things, like getting fed, happen in this chair, so he associates it as a place of security and goodness," says Ordile.

The next challenge: crate training. Gradually, Ordile introduced the pet carrier to Morris in the house, in her parked car

Got a Hollywood Kitty in the Making?

Each year, members of the Friskies Cat Team and Friskies animal trainers make special appearances at more than twenty cities in the United States to demonstrate how owners can train their cats.

Morris and celebrity cats from Friskies commercials display their talents at these shows. See page 195 for contact information.

in the garage, on trips around the block, and finally longer distances, so Morris got used to the movement and rocking.

"I train all my animals to go into their carriers when I open the door. That's their cue to go inside," she says. "They all know that going into the carrier means a positive trip or a treat or that a toy awaits them."

Over the course of the next two months, Morris learned how to make waving motions with his front paw, sit up, roll over, and stay calm in strange places like airplanes, television studios, and inside limousines.

"We made the cross-country airplane trip from Los Angeles to New York to present a $10,000 check from Heinz to the ASPCA [American Society for the Prevention of Cruelty to Animals] for his national debut and he was great," says Ordile. "This was his first time seeing taxi cabs, hearing them honking, riding in a limo, and staying in a hotel. He was more curious than scared."

Ordile's best advice to owners hoping to train their cats: practice patience and consistency and you will be able to teach any cat at any age a new trick.

Kitty Caper: Meet the Boogie-Board-Riding Foursome

Got a water-loving kitty that likes to ride in the car and explore new places? He may be a perfect candidate to surf shoreline waves or at least swim a few laps in your bathtub.

Think of kitty surfing as an extreme sport for felines, a trick reserved for the very brave challenge-seekers. Hector Castaner, of Miami, is fortunate to have not one but four cats that love this sport. He discovered that his cats like water after they would repeatedly join him in the shower stall.

Hector's feline friends — Flame, Flashback, Buster, and Stormy — are local celebrities at a beach in Key Biscayne, Florida. One or two Saturday mornings a month, Hector loads up the cats in his sports utility vehicle and heads for the warm, beckoning waters of the Atlantic Ocean.

He selects a quiet place on the shore where there aren't a lot of people. Then, he walks his leashed cats down toward the shore. He takes one cat out into the water at a time while the other three watch the action from the shore under the supervision of one of Hector's friends.

With their incredible sense of balance, his cats are born boogie-board riders. Hector swims alongside the board as each of his cats conquers the mild waves. If one of his cat should lose his balance and falls into the water, Hector is right there to scoop him up, offer encouragement, and put him back on the board.

Once back on shore, Hector rinses the salt water from his cats, towels them dry, and rewards them with treats before heading back in the car for the short trek home.

"My cats love the beach, and they purr when I start drying them," says Hector. "They trust me and know that nothing bad will happen when they are with me."

Airlines and Pet Policies

More than 500,000 pets travel by airplane each year, according to the Airline Transportation Association. Here is a rundown of the pet-travel policies among the major airlines serving the United States:

Airline	Reservations
Alaskan (affiliate: Horizon Air) 800-426-0333 (reservations)	At least 24 hours in advance; pets may travel as carry-on, checked-in baggage, or cargo
American 800-433-7300 (reservations)	At least 24 hours in advance
Continental 800-575-3335 (Live-Animal Desk)	At least 24 hours in advance
Delta (affiliates: ASA, Comair, SkyWest, Trans State Shuttle) 800-221-1212 (reservations)	Advance; acceptance first come, first served
Northwest 888-692-4738 (Priority Pet Center Information Line)	At least 24 hours in advance
TWA 800-221-2000 (reservations)	At least 24 hours in advance
United 800-241-6522 (reservations)	Advance; reconfirmed 24 to 48 hours before departure
US Airways 800-428-4322 (reservations)	Advance; acceptance first come, first served

*Health certificates must be completed by a licensed veterinarian.

Health Certificate*	Flying Restrictions
Within 30 days of travel	Pets checked-in must fly on same flight as owner and be at gate within one hour of departure
Within 10 days of travel	No pets accepted as checked baggage from May 15 through September 15
Within 10 days of travel	
Within 10 days of travel	Limited to one carry-on pet in first class, one in business class, and two in the main cabin per flight. No pets accepted as checked bags during June, July, or August or when temperatures are forecasted to be above 85°F at all points of travel; for snub-nosed breeds, the temperature must be below 70°F.
Within 10 days of travel	No pets may fly when temperatures are forecasted to be below 10°F or above 75°F for snub-nosed breeds. Any pet/kennel combo exceeding 150 pounds must be shipped as freight.
Within 10 days of travel	No pets are accepted as checked baggage from June 15 through September 7 and any time when the temperature if forecast to exceed 85°F.
Within 10 days of travel if pet travels with you; within 30 days if pet travels as cargo	Cabin space limited to one pet in first class and two pets in coach per flight. No pets may travel as checked baggage during extreme hot or cold temperatures.
Within 10 days of travel	

Resources

Emergencies

ASPCA Animal Poison Control Center
1717 South Philo Road, Suite 36
Urbana, IL 61802
Phone: 217-337-5030
Hotline: 888-4ANI-HELP (888-462-4435)
Web site: www.napcc.aspca.org/
Professionals staff the hotline, which operates 24 hours, 7 days per week; $45 fee per case.

PetFinders
Phone: 800-666-5678.
Web site: www.petfinder.org
Based in Athol, New York; a national service helping owners find lost pets.

Poisonous Plants Home Page
Cornell University
Web site: www.ansci.cornell.edu/plants/plants.html/
Based in Ithaca, New York; veterinary school providing information on plants poisonous to pets.

United Animal Nations
Emergency Animal Rescue Service
P.O. Box 188890
Sacramento, CA 95818
Phone: 916-429-2457
Web site: www.uan.org/ears/
Nonprofit organization providing emergency care for pets during disasters.

Medical Associations

American Animal Hospital Association
P.O. Box 150899
Denver, CO 80215
Phone: 303-986-2800
Web site: www.healthypet.com
Organization of veterinary hospitals providing health care information and referrals on companion animals.

American Association of Feline Practitioners
530 Church Street, Suite 700
Nashville, TN 37219
Phone: 615-259-7788
Web site: www.avma.org/aafp/
Represents veterinarians who specialize in providing medical care to cats.

American Veterinary Medical Association
1931 N. Meacham Road, Suite 100
Schaumburg, IL 60173
Phone: 847-925-8070
Web site: www.avma.org
Association of licensed veterinarians working on issues important to the veterinary profession and pet owners.

Alternative Therapies

Academy of Veterinary Homeopathy
751 NE 168th Street
North Miami, FL 33162
Phone: 305-652-5372
Web site: www.acadvethom.org
Establishes guidelines for postdoctoral education and certificate training in homeopathy medicine for practicing veterinarians.

American Holistic Veterinary Medical Association
2214 Old Emmorton Road
Bel Air, MD 21015
Phone: 410-569-0795
Web site: www.altvetmed.com
Founded in 1982; provides a referral service of its active members to the public and offers continuing education seminars for practicing veterinarians.

International Veterinary Acupuncture Society
P.O. Box 271395
Fort Collins, CO 80527
Phone: 970-266-0666
Web site: www.ivas.org
Established in 1974; offers extensive certificate training programs to its 1,200 members.

Humane Societies

Alley Cat Allies
1801 Belmont Road NW, Suite 201
Washington, DC 20009
Phone: 202-667-3630
Web site: www.alleycat.org
Provides education on how to humanely trap, spay, and neuter stray and feral cats.

American Humane Association
63 Inverness Drive East
Englewood, CO 80112
Phone: 303-925-9453
Web site: www.americanhumane.org
Founded in 1877; provides education on pet care to owners, shelters, and the veterinary community.

American Society for the Prevention of Cruelty to Animals (ASPCA)
424 East 92nd Street
New York, NY 10128
Phone: 212-876-7700
Web site: www.aspca.org
Nonprofit organization championing the humane treatment of animals.

Delta Society
289 Perimeter Road East
Renton, WA 98055
Phone: 800-869-6898
Web site: www.deltasociety.org
Provides medical research on benefits of companion animals and trains therapy animals.

Dumb Friends League
2080 S. Quebec Street
Denver, CO 80231
Phone: 303-696-4941
Web site: www.ddfl.org
Founded in 1910; one of oldest and largest humane animal shelters in the United States.

Doris Day Animal League
227 Massachusetts Avenue, NE,
Suite 100
Washington, DC 20002
Phone: 202-546-1761
Web site: www.ddal.org
Named in honor of actress Doris Day; nonprofit, national, citizen's lobbying organization focusing on issues involving the humane treatment of animals.

Humane Society of the United States
2100 L Street, NW
Washington, DC 20037
Phone: 301-258-3072
Web site: www.hsus.org
Founded in 1954; world's largest animal protection organization.

Morris Animal Foundation
45 Inverness Drive East
Englewood, CO 80112
Phone: 800-243-2345
Web site: www.morrisanimalfoundation.org
Private organization funding humane health studies of companion animals.

Cat Associations

American Association of Cat Enthusiasts
P.O. Box 213
Pinebrook, NJ 07058
Phone: 201-335-6717
Web site: www.aaceinc.org
Organization for cat breeders.

American Cat Fanciers' Association
P.O. Box 203
Point Lookout, MO 65726
Phone: 417-334-5430
Web site: www.acfacat.com
Founded in 1955; one of the world's largest cat organizations; strives to promote the welfare, education, knowledge, and interest in all domesticated, purebred, and nonpurebred cats, to breeders, owners, and exhibitors of cats.

Cat Fanciers' Association
P.O. Box 1005
Manasquan, NJ 08736
Phone: 732-528-9797
Web site: www.cfainc.org
World's largest registry of pedigreed cats; champions the cause of all cats and promotes responsible breeding.

Cat Fanciers' Federation
P.O. Box 661
Gratis, OH 45330
Phone: 513-787-9009
Web site: www.cffinc.org
Feline registry with clubs and judges in the Midwest and eastern United States.

The International Cat Association
P.O. Box 2684
Harlingen, TX 78551
Phone: 210-428-8046
Web site: www.tica.org
One of the world's largest registries of purebred and household cats; sanctions cat shows and promotes the welfare of cats.

Publications

***Cat Fancy* Magazine**
P.O. Box 6050
Mission Viejo, CA 92690
Phone: 949-855-8822
Web site: www.animalnetwork.com
Monthly publication geared to cat owners that covers health care, cat behavior, and breed profiles.

***CatNip* Newsletter**
Tufts University School of Medicine
P.O. Box 420014
Palm Coast, FL 32142
Web site: www.tufts.edu/vet/publications/catnip/
Monthly newsletter for cat owners and enthusiasts that is under the auspices of Tufts University School of Veterinary Medicine.

***Cats* Magazine**
Primedia Special Interests
260 Madison Avenue, 8th Floor
New York, NY 10016
Web site: www.catsmag.com
Monthly publication for cat owners and enthusiasts.

Cat Watch Newsletter
Cornell University College of Veterinary Medicine
P.O. Box 420235
Palm Coast, FL 32142
Web site: www.vet.cornell.edu/publicresources/cat.htm
Monthly newspaper for cat people.

Kittens U.S.A
Fancy Publications
P.O. Box 6050
Mission Viejo, CA 92690
Phone: 949-855-8822
Web site: www.animalnetwork.com
Yearly publication on care for kittens.

Services

Friskies Talent Search
Phone: 800-725-CAT
The Friskies Cat Team and animal trainers make special appearances each year to demonstrate how owners can train their cats. If your kitty is a budding star, contact Friskies officials at the number above.

National Association of Professional Pet Sitters
6 State Road, Suite 113
Mechanicsburg, PA 17005
Phone: 800-296-PETS
Web site: www.petsitters.org
Founded in 1989; national professional organization of licensed pet sitters promoting in-home pet care and referrals for cat owners.

Pet Sitters International
418 East King Street
King, NC 27021-9163
Phone: 336-983-9222
Web site: www.petsit.com
Educational organization providing a network for professional pet sitters and referrals for cat owners.

Spay/USA
2261 Broadbridge Avenue
Stratford, CT 06614
Phone: 800-248-SPAY
Web site: www.spayusa.org
National clearinghouse for all spay and neuter programs. Open weekdays 9:30 A.M. to 4:30 P.M. E.S.T. More than 8,000 veterinarians in 950 programs participate.

Tattoo-A-Pet
6571 SW 20th Ct
Ft. Lauderdale, FL 33317
Phone: 800-828-8667
Web site: www.tattoo-a-pet.com
Established in 1972; world's largest network of pet protection registration and recovery.

VetQuest
Pet Care Forum, PMB 106-131
141 W. Covell Blvd.
Davis, CA 95616
Phone: 800-700-4636
Web site: www.vetquest.com
Veterinary referral service.

Index

Other Storey Titles You Will Enjoy

Dr. Kidd's Guide to Herbal Cat Care by Randy Kidd, D.V.M., Ph.D. This comprehensive guide to herbal health care for cats is written by a practicing holistic veterinarian. 208 pages. ISBN 1-58017-188-5.

50 Simple Ways to Pamper Your Cat by Arden Moore. Pet expert Arden Moore takes a lighthearted look at pampering the all-deserving feline. Filled with tips, techniques, and recipes, this book will be appreciated by cat enthusiasts — and cats — everywhere. 144 pages. ISBN 1-58017-311-X.

Dr. Kidd's Guide to Herbal Dog Care by Randy Kidd, D.V.M., Ph.D. Readers learn how to use natural methods, such as herbal remedies and massage, to help maintain their dog's health. 208 pages. ISBN 1-58017-189-3.

50 Simple Ways to Pamper Your Dog by Arden Moore. Who deserves pampering more than man's (and woman's) best friend? Arden Moore makes it fun to shower a pooch with love, affection, home-baked treats, and all the rewards that your loyal canine deserves. 144 pages. ISBN 1-58017-310-1.

The Puppy Owner's Manual by Diana Delmar. In a fun question-and-answer format, Diana Delmar offers humane, common sense solutions for chewing, barking, grooming, eating, house training, and other puppy problems. 192 pages. ISBN 1-58017-401-9.

These and other books from Storey Publishing are available wherever quality books are sold or by calling 1-800-441-5700. Visit us at www.storey.com.